Terrorism, Security, and Computation

Series Editor
V. S. Subrahmanian
Department of Computer Science and Institute for Security, Technology and
Society, Dartmouth College, Hanover, NH, USA

The purpose of the Computation and International Security book series is to establish the state of the art and set the course for future research in computational approaches to international security. The scope of this series is broad and aims to look at computational research that addresses topics in counter-terrorism, counterdrug, transnational crime, homeland security, cyber-crime, public policy, international conflict, and stability of nations. Computational research areas that interact with these topics include (but are not restricted to) research in databases, machine learning, data mining, planning, artificial intelligence, operations research, mathematics, network analysis, social networks, computer vision, computer security, biometrics, forecasting, and statistical modeling. The series serves as a central source of reference for information and communications technology that addresses topics related to international security. The series aims to publish thorough and cohesive studies on specific topics in international security that have a computational and/or mathematical theme, as well as works that are larger in scope than survey articles and that will contain more detailed background information. The series also provides a single point of coverage of advanced and timely topics and a forum for topics that may not have reached a level of maturity to warrant a comprehensive textbook

More information about this series at http://www.springer.com/series/11955

V. S. Subrahmanian • Judee K. Burgoon
Norah E. Dunbar

Editors

Detecting Trust and Deception in Group Interaction

Editors
V. S. Subrahmanian
Department of Computer Science
Dartmouth College
Hanover, NH, USA

Judee K. Burgoon
University of Arizona
Tucson, AZ, USA

Norah E. Dunbar
Department of Communication
University of California, Santa Barbara
Santa Barbara, CA, USA

ISSN 2197-8778   ISSN 2197-8786   (electronic)
Terrorism, Security, and Computation
ISBN 978-3-030-54385-3   ISBN 978-3-030-54383-9   (eBook)
https://doi.org/10.1007/978-3-030-54383-9

This Springer imprint is published by the registered company Springer Nature Switzerland AG
The registered company address is: Gewerbestrasse 11, 6330 Cham, Switzerland

Preface

With the rapid increase in the amount of video available online, there is growing interest in analyzing the content of such videos.

Social scientists have an interest in analyzing videos of a group of people interacting together in order to better understand the factors that lead to a host of behaviors exhibited in group interactions. In particular, social scientists would like to understand facial and vocal cues that are linked to answers to questions such as: Does person X like person Y? Does person X trust person Y? Is person X more dominant than person Y? Is person X being deceptive? The use of carefully curated, data-driven studies such as these are incredibly helpful in validating existing theories linking facial and vocal cues to such behaviors.

At the same time, numerous corporations and governments have an interest in the automated analysis of videos of group interactions. Recent advances in machine learning offer the promise of building end-to-end systems that analyze video and automatically predict the answers to questions such as those listed above. We note that prediction of a behavioral event and understanding the factors that are cues to the occurrence of a behavior are not the same thing.

This edited book brings together a series of chapters that build on a novel dataset created by the authors and their students and collaborators. By focusing on carefully collected video of a popular face-to-face game called *The Resistance*, we present a comprehensive overview of the power of both social science theory and computational modeling in understanding and predicting behavior during interactions among a group of people.

We are grateful to numerous people who have helped produce this book. First, we would like to express our gratitude to the Army Research Office for funding much of the work reported in this book under Grant W911NF-16-1-0342. In particular, we would like to thank Dr. Purush Iyer, Dr. Edward Palozzolo, and Dr. Lisa Troyer for their very strong support to this work, along with a continuous stream of valuable comments, encouragement, and advice. Thanks are also due to Dr. Addison Bohannon, Dr. Liz Bowman, Dr. Javier Garcia, and Dr. Jean Vettel for sharing a host of interesting and useful thoughts. Of course, we are grateful to the PIs on this project from numerous institutions: Profs. Larry Davis, Jure Leskovec,

Miriam Metzger, and Jay Nunamaker. We are also deeply grateful to Chongyang Bai for his hard work in formatting the chapters and helping to pull all the chapters submitted into a single document. Finally, we express our sincere gratitude to many of the students and postdocs who helped shape this effort and generate the results presented in this book.

Hanover, NH, USA   V. S. Subrahmanian
Tucson, AZ, USA   Judee K. Burgoon
Santa Barbara, CA, USA   Norah E. Dunbar

Feb 27, 2020

Contents

Contributors

Chongyang Bai Dartmouth College, Hanover, NH, USA

Maksim Bolonkin Dartmouth College, Hanover, NH, USA

Judee K. Burgoon University of Arizona, Tucson, AZ, USA

Xunyu Chen University of Arizona, Tucson, AZ, USA

Bradley Dorn University of Arizona, Tucson, AZ, USA

Norah E. Dunbar University of California, Santa Barbara, Santa Barbara, CA, USA

Becky Ford University of California, Santa Barbara, Santa Barbara, CA, USA

Saiying (Tina) Ge University of Arizona, Tucson, AZ, USA

Matt Giles University of California, Santa Barbara, Santa Barbara, CA, USA

Mohemmad Hansia University of California, Santa Barbara, Santa Barbara, CA, USA

Joel Helquist Utah Valley University, Orem, UT, USA

Matthew L. Jensen University of Oklahoma, Norman, OK, USA

Srijan Kumar Stanford University, Stanford, CA, USA

Jure Leskovec Stanford University, Stanford, CA, USA

Dimitris Metaxas Rutgers University, New Brunswick, NJ, USA

Miriam Metzger University of California, Santa Barbara, Santa Barbara, CA, USA

Jay F. Nunamaker University of Arizona, Tucson, AZ, USA

Steven J. Pentland Boise State University, Boise, ID, USA

Lee Spitzley University at Albany, Albany, NY, USA

V. S. Subrahmanian Dartmouth College, Hanover, NH, USA

Brad Walls University of Arizona, Tucson, AZ, USA

Xinran (Rebecca) Wang University of Arizona, Tucson, AZ, USA

Lezi Wang Rutgers University, New Brunswick, NJ, USA

Karl Wiers University of Arizona, Tucson, AZ, USA

Part I
Theory Underlying Investigating Deception in Groups

Chapter 1
Prelude: Relational Communication and the Link to Deception

Judee K. Burgoon

In 1984, when Jerry Hale and I published our first article on "The Topoi of Relational Communication" (Burgoon and Hale 1984), it had been 3 years in the making, 3 years trying to synthesize the perspectives of the various social science rivulets feeding into our conceptual stream of relational communication themes. Some scholars took issue with our advocacy of 12 interdependent themes, arguing that the two superordinate dimensions of dominance-submission and affection-hostility held longstanding status in the world of interpersonal relationships and all the other themes were merely auxiliaries, while others took issue with our methodology for uncovering meaningful message themes (Burgoon and Hale 1987). But our objective had been to highlight those "stock" topoi—ones devoid of specific content but relevant to a dyad member's status vis à vis another—that communicate how members feel about one another or could be used by a third party to characterize the relationship: they trust each other implicitly (trust-distrust), she greets him with warmth and inclusiveness (affection-hostility), their understanding of one another is a mile wide and an inch deep (depth-superficiality), she expects her employees to show respectful decorum with her (formality-informality) and they in turn regard her communication with them as cold and detached (involvement-detachment), he becomes distressed whenever his boss enters the room (emotional arousal-relaxation) and his posture shrinks to become more diminutive and appeasing (dominance-submission).

These dimensions, and several others, each carry nuanced meaning that "speaks," mostly nonverbally, to how participants in a relationship regard one another, the relationship itself and themselves within the relationship. Each theme has an amalgam of nonverbal and linguistic signals that are non-redundant with those of other

J. K. Burgoon (✉)
University of Arizona, Tucson, AZ, USA
e-mail: judee@email.arizona.edu

© Springer Nature Switzerland AG 2021
V. S. Subrahmanian et al. (eds.), *Detecting Trust and Deception in Group Interaction*, Terrorism, Security, and Computation,
https://doi.org/10.1007/978-3-030-54383-9_1

themes. If they were redundant, they would lose their unique meanings and could be eliminated.

Application to Deceptive Group Interaction

Within the current project, the purpose of the relational dimensions is twofold: (1) to gauge how group members regard one another by observing their displayed relational messages rather than questioning them directly and (2) to gauge whether relational messages predict who is deceptive and who is not. In the former case, are relational messages of dominance, arousal and trust evident in how group members behave? In the latter case, would it be beneficial to measure the behavioral indicators directly or to fuse them into their constituent relational message themes to predict veracity?

In a previous deception experiment employing a mock theft, Jensen et al. (2008) applied a Brunswikian lens model to "identify configurations of micro-level deception cues that predict mid-level percepts which in turn predict attributions" (p. 428). Analyzed were measures of tension, arousal and involvement. Due to low intraclass reliability for the measure of dominance, it was omitted from the analysis. When multiple predictors were included in the model, arousal and involvement emerged as predictors of deception. Deceivers (those who committed a theft of a wallet from a classroom) were more aroused and less involved than truth tellers (those who were innocent bystanders in the classroom during the theft). When single predictors were used, tension was also negatively associated with honesty; more tension was associated with dishonesty or deception. Thus, relational expressions of arousal, nonrelaxation and involvement were relevant to distinguishing truthful from lying interactants.

In the context of the current experiment, relational communication plays a role in two respects. Here we provide a brief overview of how relational message themes gauge interpersonal relations across time and how they predict team member veracity. The scenario in use is a game in which players designated as Villagers attempt to thwart infiltration by players designated as Spies. The Villagers are assumed to be truth tellers. Spies are assumed to be deceivers. Through successive rounds of missions, Villagers vote for missions to succeed and Spies vote for missions to fail. Along the way, team members must vote for leaders of each mission and must approve the team members who are proposed to be on a mission team.

Relational Communication Hypotheses

Relational messages come into play (1) as covert communication among Spies, who know each other's identity and must decide how to collude with one another surreptitiously to make missions fail, (2) as overt messages among Villagers who must make decisions about who to select as mission team leaders and team members, and (3) finally, as implicit communication between Spies and Villagers, positioning themselves along dominance-submission and trust-distrust continua as they negotiate mission team composition and leadership.

Spies, who must be deceptive if they are going to sabotage the missions and win the game, would seem to have a few alternatives at their disposal. One is to express to other team members that they are involved in the game and trustworthy so that they are selected as leaders and team members. Another is to convey that they are calm, relaxed, and not distressed, on the assumption that displays of distress might be read as signs of nervousness and deception.

Villagers in turn may be tentatively putting out feelers to ascertain who to trust and who can help them win the game. Conversely, they may be on the look-out for signals that feed suspicion and distrust as they try to discern who may sabotage the game. Apart from overt accusations of those they believe are spies, players must rely on relational communication to compete and complete the game.

We expected that Spies might opt initially to be passive, "hiding in the weeds" as it were, to keep their identity concealed but they might increasingly engage in "persuasive deception" (Dunbar et al. 2014) over time to earn other team members' support. We also expected that Spies might betray some nervousness, compared to the Villagers. Finally, we expected that Villagers would be seen as more trustworthy than Spies, although if Spies succeeded in their persuasive efforts, they might gain as much trust as Villagers.

Method

To assess how relational messages might be utilized in the games, the players were asked after every other round to complete self-report measures of how nervous (aroused) or relaxed, dominant or submissive and trustworthy or suspicious the other players were. Additionally, ratings were collected after a beginning ice breaker phase of the game, which served as a baseline for comparison. Scales ranged from 1 (not at all) to 5 (very). Only Villagers' ratings were considered since Spies' ratings would be contaminated by their knowledge of one another's role.

Results

A repeated measures regression analysis was conducted across the rated rounds of the game. The results for nervousness across four periods of ratings, beginning with the baseline and ending with the final ratings, produced a significant main effect, $F(3,996) = 5.23, p = .001$, partial $\eta^2 = .016$, and an interaction between nervousness and game role, $F(3,996) = 2.75, p = .041$, partial $\eta^2 = .008$. Whereas Spies maintained the same degree of nervousness they displayed at the outset of the game,

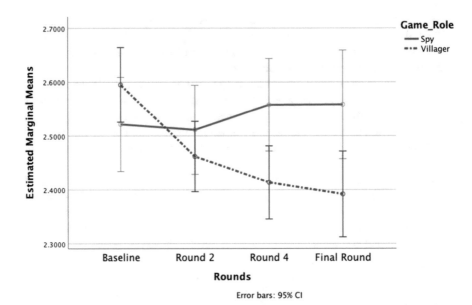

Fig. 1.1 Villager and Spy nervousness over time

Villagers became increasingly relaxed (see Fig. 1.1). Thus, an astute observer might have noticed that some of the players in the group were not as calm and collected as the majority. Villagers might have made use of this signal to identify Spies, and Spies might have used this subtle difference to confirm the identity of other Spies, although the difference between Spies and Villagers was rather slight. Thus, it wasn't that Spies signaled *more* nervousness but rather that they failed to show the increased relaxation that characterized the truth-telling Villagers. They might be credited with communicating the same degree of arousal that they had at the start of the game, before roles were even known, but they still set themselves apart from the calmer Villagers.

The same repeated measures analysis on ratings of trust produced main effects for game role, $F(1,332) = 104.22, p < .001$, partial $\eta^2 = .239$, trust ratings across time $F(3,996) = 127.67, p < .001$, partial $\eta^2 = .278$, and the interaction between game role and trust, $F(3,996) = 47.00, p < .001$, partial $\eta^2 = .124$ (see Fig. 1.2). Even though Villagers did not know who the Spies were, their ratings showed they trusted the Spies less, and those ratings continued to decline over the course of the game. Ratings of Villagers, by contrast, remained higher and showed an upswing over time, although never returning to their baseline level, possibly because suspicion about the presence of Spies tempered their judgment somewhat. When only four rounds were included in the analysis, results were even more striking. Thus, participants' perceived trustworthiness of one another was a good barometer of who actually could be trusted.

Finally, results on dominance largely paralleled those for trust. Because dominance is analyzed extensively in other chapters in this volume, they are omitted here.

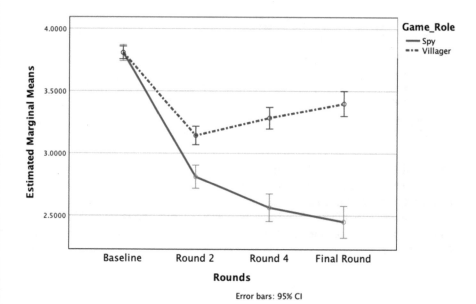

Fig. 1.2 Mean trust ratings by game role and rounds

Table 1.1 Tests of equality of group means between spies and villagers

	Wilks' Lambda	F	df1	df2	Significance
Trust	.848	123.197	1	687	<.0001
Dominance	.964	25.505	1	687	<.0001
Arousal	.993	5.173	1	687	.023

The paramount objective in the SCAN project is to identify deceivers. Can the relational messages offer tacit indications of who is truthful and who is not? An initial analysis with location included as a covariate to capture possible cultural differences showed location to be nonsignificant. However, all three relational message dimensions independently distinguished between Spies (deception) and Villagers (truth), as shown in Table 1.1.

To determine the ability of the relational messages to predict who were Spies and who were Villagers, we conducted a multiple discriminant analysis. The first and most powerful predictor was trust: Trust was much higher for truth tellers than deceivers, indicating that relational messages were a good signal of who was actually a deceiver and who, not (see Table 1.2). Only one predictor remained in the final model, Wilks' $R_c = .39$. The cross-validated classification matrix showed the model accurately identified Villagers at 79% but only identified Spies at 55% accuracy. These results might suggest that other data would be needed to accurately identify the Spies.

However, by taking the dynamics of judgments into account and including measurements of the relational messages for each round, a different picture emerges. If

Table 1.2 Discriminant analysis distinguishing spies (deceivers) from truth tellers

Step	Entered	Wilks' Lambda statistic	df1	d2	d3	Exact F	df1	df2	Significance
1	Final trust rating	.828	1	1	630	131.16	1	630	<.0001
2	Baseline dominance	.820	2	1	630	69.02	2	629	<.0001
3	Round4 trust	.812	3	1	630	48.33	3	628	<.0001
4	Final dominance	.807	4	1	630	37.55	4	627	<.0001

ratings from participants who completed at least three rounds are included in the analysis, better predictions develop. Four variables are in the final model: last round trust, second round trust, last round dominance and baseline dominance. The four-variable model was highly significant, Wilks' R_c = .553. The cross-validated classification matrix showed the model accurately identified Villagers at 81% and Spies at 65% accuracy.

Discussion

These results underscore the importance of more granular, temporal measurement. The variance accounted for, and the accuracy of distinguishing truth tellers from deceivers, show a robust assessment of deception that is evident from relational communication alone. Measurement is also important at each juncture of interaction. That is, impressions at different stages of the group process add information to the ability to predict veracity. Interestingly, at none of the junctures does the traditional signal of arousal enter the model. In other words, even though nervousness discriminated between Spies and Villagers, considerations other than nervousness are even more important indicators of truth and deception.

Anytime humans are involved in an activity involving social interaction, how they regard one another and their interpersonal relationship can create a fluid situation that "greases the skids," or one that creates barriers to forward progress. Ironically, in the case of deception, what is needed is knowledge of the barriers that lead to thoughtful scrutiny of others rather than facile acceptance. It may be that relational communication becomes the leading edge in assessing the truthfulness or deceptiveness of others. In the chapters that follow in this volume, many ways of gauging dominance, arousal, and trust are analyzed along with their ability to predict deception. The overriding message is that relational messages are an important signpost in group interaction of the interpersonal relationships among group members and can alert one to suspicions and distrust even when such sentiments are not spoken aloud.

Acknowledgement We are grateful to the Army Research Office for funding much of the work reported in this book under Grant W911NF-16-1-0342. The views and conclusions contained in this document are those of the authors and should not be interpreted as representing the official policies, either expressed or implied, of the Army Research Office or the U.S. Government. The U.S. Government is authorized to reproduce and distribute reprints for Government purposes notwithstanding any copyright notation herein.

References

Burgoon, J. K., & Hale, J. L. (1984). The fundamental topoi of relational communication. *Communication Monographs, 51*, 193–214.

Burgoon, J. K., & Hale, J. L. (1987). Validation and measurement of the fundamental themes of relational communication. *Communication Monographs, 54*, 19–41.

Dunbar, N. E., Jensen, M. L., Bessabarova, E., Burgoon, J. K., Bernard, D. R., Robertson, K. J., Kelley, K. M., Adame, B., & Eckstein, J. M. (2014). Empowered by persuasive deception: The effects of power and deception on interactional dominance, credibility, and decision-making. *Communication Research, 41*, 852–876.

Jensen, M. L., Meservy, T. O., Burgoon, J. K., & Nunamaker, J. F., Jr. (2008). Video-based deception detection. In H. Chen & C. C. Yang (Eds.), *Intelligence and security informatics: Techniques and applications* (pp. 425–441). Berlin: Springer.

Chapter 2
An Integrated Spiral Model of Trust

Judee K. Burgoon, Norah E. Dunbar, and Matthew L. Jensen

When humans are gathered together for work or play, the first decision they must make is whether others are friend or foe. Our emotional antennae immediately gauge whether others are equal in power or not, whether they appear to be likeable or to be feared, whether they are like one another or different. And most importantly, whether they can be trusted or not. These judgments drive the first communication exchanges humans have, and most occur implicitly, through nonverbal rather than verbal means (Burgoon and Hale 1984).

When people come together in groups, these decisions become multiplicative. They must make multiple judgments at once, and one of the most critically important ones to render is trust. When we trust one another, we can develop bonds with our colleagues, friends, family members, relationship partners, and strangers. We feel emotionally secure and confident, resolve conflicts equitably, and work collaboratively to solve problems. When we do not trust others, we feel suspicious that they want to hurt us, we are reluctant to take risks, and we feel unsupported and alone. Although much of the literature on trust has focused either on romantic relationships (e.g. Kim et al. 2015) or business negotiations (e.g. Koeszegi 2004), we believe that trust, as Kim et al. eloquently stated, is a "central component of nearly all good, well-functioning relationships because it allows individuals to pursue their loftiest hopes without being impeded by their deepest anxieties" (p. 522).

J. K. Burgoon (✉)
University of Arizona, Tucson, AZ, USA
e-mail: judee@email.arizona.edu

N. E. Dunbar
University of California, Santa Barbara, Santa Barbara, CA, USA

M. L. Jensen
University of Oklahoma, Norman, OK, USA

© Springer Nature Switzerland AG 2021   11
V. S. Subrahmanian et al. (eds.), *Detecting Trust and Deception in Group Interaction*, Terrorism, Security, and Computation,
https://doi.org/10.1007/978-3-030-54383-9_2

Definitions of the Nature of Trust

Regardless of discipline, most scholars conceptualize trust as entailing some level of risk, uncertainty, or willingness to be vulnerable, and that it creates an *expectancy* about future behavior since one must assume that a person, group, or organization will behave in a particular way (Lewicki et al. 1998; Rousseau et al. 1998). Typically, trust is built slowly over time as judgments about past behavior are evaluated and the costs and benefits (or risks and rewards) of such future behavior are cognitively assessed (Robert et al. 2009). In long-term relationships, theorists have investigated trust in the initial phases of relationship building (Taylor and Altman 1987), in its maintenance (Rempel et al. 2001), and in its dissolution (Sagarin et al. 1998). Thus, the traditional or developmental view of trust would predict low levels of initial trust because team members have little past history, may not share common cultures, and have few personal observations on which to assess risk (Robert et al. 2009). However, even when relationships are relatively brief and temporary, interactants rely on contextual cues and initial expectations to create expectancies about trust (called "swift trust;" Adler 2007). Often, swift trust is created based on characteristics prior to any knowledge of the others' actual behavior. Robert et al. (2009) argue that swift trust is based on factors other than past behavior such as one's role, disposition, sociological category like gender and culture, and third-party recommendations. Zero-history teams (such as cockpit crews or investigative task forces) consist of members with diverse skills, have a limited history of working together, and often have little prospect of working together again in the future, which make it difficult to build trust. The tight deadlines under which these teams work leave little time for relationship building, but trust must still be developed in order to be an effective team (Jarvenpaa et al. 1998), especially among teams that are geographically dispersed and must accomplish their tasks rapidly (Iacono and Weisband 1997; Meyerson et al. 1996). Because the very concept of social organization relies on reciprocal trust, good will, and cooperation (Gouldner 1960; Grice 1989), people are usually inclined to give one another the benefit of the doubt, to view each other as truthful and trustworthy. This "truth bias" is bolstered by the human tendency to regard all incoming information as truthful and, only after digesting and reflecting upon it, to entertain the possibility that it may be false (Gilbert et al. 1990). Thus, the default orientation in most "cooperative" situations should be toward mutual trust.

Nevertheless, trust is not a given. It depends fundamentally on the interpersonal relationships among individuals. People are predisposed to trust others whom they know well because they have a basis for assessing each other's expertise, sound judgment, honesty, reliability, poise, and so forth. Early in relationships and among previously unacquainted team members, trust is provisional and probationary, and it is inextricably linked to the communication that ensues (Hinsz et al. 1997). Moreover, if relationships are thought to be adversarial, or when group members come from diverse backgrounds, the truth bias is attenuated and trust must be built (Foddy et al. 2009; Grice 1989; Lewicki et al. 1998). This is particularly true when an ingroup-outgroup divide exists, as may be the case between people of different cultural backgrounds, clans, organizations, or even genders (Yuki et al. 2005). In such cases, a communication perspective argues that the dynamics of the interactions among team members will determine the trajectory of trust. Critically important is the extent to which participants are able to adapt to one another's communication patterns and to achieve a coordinated, synchronized interaction style that creates perceptions of common ground and understanding (Burgoon et al. 1995b; Tickle-Degnen and Rosenthal 1990). Although the content of what is spoken obviously is relevant, as team members appraise one another's claims and proffers, how the discourse is transacted may be far more important in creating perceptions of rapport, genuineness, positive motives, and trustworthiness (Duggan and Parrott 2001; Fiksdal 1988; Heintzman et al. 1993).

Psychological and Sociological Perspectives of Trust

Psychologists and sociologists often implicitly assume that communication is involved in the trust-development process but often overlook the critical mediating role of the interaction itself (Jung and Avolio 2000; see Peters and Kashima 2007; Silvester et al. 2007; Van Overwalle and Heylighen 2006; Buchan et al. 2006). Researchers often disregard the joint and emergent social processes that are critically important to determining the trajectory of trust. For example, in many studies of trust like Ho and Weigelt's (2005) study of trust building among strangers, pre- and post-interactional factors such as social uncertainty and exchange outcomes, are posited to have an effect on the process of trust building. Participants play a trust game in which points are earned based on decisions made during the game and the outcomes are analyzed. Ho and Weigelt argue that their trust game allows subjects to reveal their trustworthiness by choosing to share their social gains with others. The authors examine the monetary payoffs as their primary measure of trust. In contrast, communication scholars posit a direct relationship between interactive communication processes and relationship development as well as interaction outcome evaluations (Burgoon et al. 1995a; Dunbar et al. 2014; Manusov et al. 1997). In our view, the verbal and nonverbal messages exchanged, the subtle cues that partners give off as they coordinate their interaction, and the interpersonal relationship that develops (even in temporary teams) are crucial to understanding the nature of trust. Although trust games (e.g. Cook et al. 2009) offer objective, easily measured outcomes, they offer little to examine whether expectation fulfillment leads to trust and often ignore the interaction where those expectations are formed.

In addition, many current research paradigms suggest that trust changes in character over time, and that there is likely a feedback loop whereby the forms of trust are linked and build on each other as a relationship develops. One frequently-cited trust model (Mayer et al. 1995) suggests that outcomes serve as a catalyst for further trust growth or decline. Yet this explanation may be overly simplistic. First, it is not always the case that information about a partner's trustworthiness is fully available or unambiguous. Second, regardless of whether such outcomes are clear or ambiguous, their impact on the future development or decay of trust is likely far from direct. Third, there are likely to be multiple feedback triggers occurring at multiple time scales during and after an interaction to influence the growth or decay of trust. From psychology and sociology we know that the level of social identity between parties is a fundamental factor in the development of trust (Kramer et al. 1996; Lewicki and Bunker 1995).

One notable social-psychological approach to trust is Simpson's (2007) "dyadic model of trust" which determines whether or not trust will result from an interaction between two interdependent parties (see Fig. 2.1). The model assumes that individuals who have certain dispositions, such as secure attachments and positive self-esteem, should be more likely to enter, transform, or occasionally create trust-diagnostic situations in their relationships. The partners must be willing to take a risk and make themselves vulnerable for the sake of a mutually-beneficial outcome. Each partner makes an independent assessment of whether their partner displays proper "transformation of motivation" in trust-diagnostic situations. In other words, does the partner make decisions that go against their own interest in favor of the best interests of the partner or the relationship? If both partners make mutually-beneficial decisions, this should generate positive patterns of attributions, emotions, and future expectancies, which in turn should enhance perceptions of trust and felt security, even if it is only temporary. Although the Simpson model does not depict any feedback loops, it is presumed that each partner's perceived degree of felt security affects future decisions about whether or not to enter, transform, or create the next trust-relevant situation.

A Communication Perspective on Trust

This brings us to our communication perspective on trust. Building on Simpson's (2007) model, we assume that trust is an *interactive* and *iterative* process of evaluations of motives that is affected by past interactions and dispositions. Within the

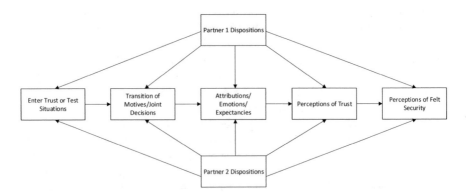

Fig. 2.1 Adapted from Simpson's (2007) dyadic model of trust

communication literature, trust is one component of credibility, a construct itself that has a venerable heritage tracing back to Aristotle's treatise on ethos (Hovland et al. 1953). Credibility is a judgment that others make about an actor. It usually includes, at minimum, the dimensions of competence (knowledge, intelligence, experience, and authoritativeness) and character (trustworthiness, reliability, and honesty) but may also include components of sociability, composure, dynamism, extroversion, empathy, and good will toward the other (Burgoon 1976; McCroskey and Young 1981; Teven and McCroskey 1997). Recent research has demonstrated that in the context of task-oriented interactions, many of these components can be fruitfully combined with each other and with other social judgments. For example, judgments related to credibility, attraction, and utility are sufficiently correlated that they can be combined into a smaller number of composite measures, all with possible relevance to trust (Burgoon et al. 1999). Through factor analysis, Burgoon et al. identified the following dimensions: (1) *trust* (perceptions of honesty, truthfulness, sincerity, and character), (2) *dependability* (perceptions of being reliable, helpful, useful, and responsible), (3) *expertise* (perceptions of competence, knowledge, and experience), (4) *sociability* (perceptions of friendliness and good will), and (5) *attraction* (desirability of another as a task partner and that person's likely contributions to task performance). One can "trust" others because they are thought to be honest, forthright individuals, or because they are dependable and helpful, or because they have the necessary knowledge and judgment to contribute to task performance, or because they are thought to be a person with others' interests at heart, or because they have performed ably on a given task and contributed to one's satisfaction with the work. In other words, all of these facets of credibility have relevance to trust.

Another way trust-distrust is conceptualized in the communication literature is as one of the fundamental dimensions along which people define and understand their interpersonal relationships (Burgoon and Hale 1984). Like power or status, trust is one of the central relational communication themes by which people express and calculate, verbally and nonverbally, the current status of their relationship with another. Like other relational messages, trust creates a frame for interpreting other messages that are exchanged. For example, in organizations, it is often the way messages are sent, especially their clarity, and a leadership style that engenders *trust*, that is of the highest importance when influencing employees' commitment to the organization (Bambacas and Patrickson 2008). When team members trust one another, messages are interpreted at face value and small disagreements or grievances may be excused or overlooked. When team members distrust one another, motives may become suspect, information may be misinterpreted, and even innocuous statements can trigger hostile reactions or noncompliance. Expressions and interpretations of trust and credibility are thus essential to the effective functioning of work groups and skillful leadership (Iacono and Weisband 1997).

When thinking about issues of trust and credibility, it is commonplace to think of trust as the desired state. However, there may be times when trust is not the goal, times when one needs to be vigilant, suspicious, or cautious toward others whose motives are not known, times when dealing with conflicting instructions requires thoughtful processing rather than blind obedience or reciprocation. Thus it becomes important to consider trust in terms of what is the desired state relative to what is actually achieved. If we cross goal states with achieved states, and for the sake of simplicity dichotomize them as trust or distrust, four kinds of situations merit attention (see Fig. 2.2): those in which trust is both desired and achieved (the prototypical case of goal achievement); those in which "distrust" (e.g., skepticism, wariness, vigilance) is desired and achieved (also a case of goal achievement); those in which trust is desired but distrust results (goal failure); and those in which distrust is desired but trust results (also goal failure). These latter circumstances seem especially worthy of focused research to understand why trust fails to develop or be sustained, or what causes undue trust.

Communicatively, trust can be revealed in a number of ways. One way is by the rapport and coordination exhibited by both parties. As the interaction begins, interactants display both intentional and unintentional nonverbal behaviors in order to express particular emotions and build rapport, liking, and trust (or divergence and distrust as required). Perceptions of intentionality appear to play a mediating role where senders who are perceived as being overly intentional in their expressivity, synchrony, and adaptations are seen as manipulative and untrustworthy. Alternatively, a degree of intentionality is required for both the successful expression of emotions and for synchronizing behaviors with an interlocutor in order to inspire attraction, rapport, and trust (Bernieri 1988; Dunbar et al. 2014; Manusov 1992).

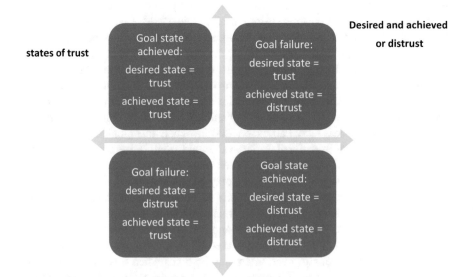

Fig. 2.2 Desired and achieved states of trust or distrust

Assumptions

Conceptualizations of trust come with implicit assumptions. We first elucidate several assumptions, describe the integration of past theories that explain trust-distrust, and then describe in detail the spiral model of trust, found in Fig. 2.3. Assumptions are beliefs accepted (or presumed) as common ground upon which the theoretical arguments are founded. Assumptions are not tested; they are taken as givens. They are the "glue" that holds testable propositions together. We articulate these assumptions before proceeding to our integration of two theoretical approaches as they pertain to trust.

1. *Trust is multidimensional.* Multiple indicators can signal the presence of trust by Other (O) toward Self (S) and vice versa.
2. *Trust derives from multiple factors.* No single, elegant model can capture all of them.
3. *Trust flows from, is sustained by, and is modified through interaction patterns.* The interaction patterns that transpire between two or more parties are among the most central factors in a communication theory of trust.
4. *Trust is best understood as a relational phenomenon.* Because interaction is jointly defined by the parties to the interaction, this requires researchers' commitment to the dyad or group as the unit of analysis.
5. *Communication is causally the most proximal variable to account for trust-distrust.* Other variables are more distal and exert their influence through the communication that transpires.

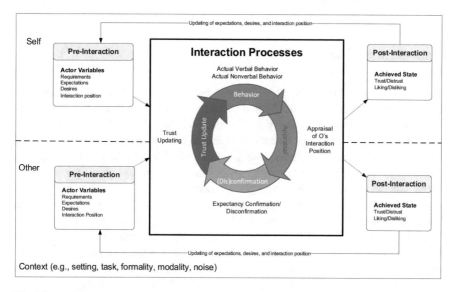

Fig. 2.3 Pre-interaction, interaction, and post-interaction stages of the spiral model of trust

6. *Because trust derives from interaction patterns, trust, like the interaction itself, is dynamic.* The implication of this assumption is that trust is not a static phenomenon. Research into trust must incorporate a temporal component into a theoretical model.

The Spiral Model of Trust: Integrating IAT and EVT

Based in part on research reviewed above from both the social-psychological and communication research literatures, we offer a process model of trust that attempts to integrate psychological factors, sociological and contextual demands, and the communicative acts that occur within dyads to predict when trust forms in human relationships. Drawing on two communication theories, *expectancy violations theory* and *interaction adaptation theory,* this process model, shown in Fig. 2.3, depicts what happens between self (S) and other (O). The process is modeled according to what precedes the interaction, the interaction itself, and where things stand after interaction. *Pre-interaction* factors include such things as a priori attitudes, personality, and social skills that guide the interaction. *Interactional* factors include how internal states relate to behavioral patterns and the extent to which behavioral patterns confirm or violate expectations. Negative violations prompt suspicion and distrust that shape the trajectory of the interaction. Positive violations promote interaction coordination and rapport. *Post-interaction* factors are task-relevant outcomes (performance, satisfaction) and social judgments (credibility, liking, and in the case of the current model, trust). The theories we describe next relate directly to these variables, offering predictions and explanations of their interrelationships.

Expectancy Violations Theory

Expectancy Violations Theory (EVT) originated as a theory about how interactants relate to one another proxemically and was subsequently expanded to include nonverbal behaviors and then verbal behavior as well (Burgoon and Hale 1988). Its key variables include expectations, arousal, appraisal, and behavioral confirmations and violations. *Expectations* (E) refer to norms and individuated anticipations of how an interlocutor will behave and communicate. Norms are group-wide socially inculcated patterns of conduct. For example, the norm of reciprocity states that "people should help those who have helped them, and people should not injure those who have helped them" (Gouldner 1960, p. 171). This norm is thought to be a fundamental and universal principle for preserving social order and underpins an expectation that others can be trusted to reciprocate kindness with kindness and to eschew harming others if unprovoked. Where people are familiar with one another, expectations are individuated; they will be tailored according to that prior knowledge.

S's behavior might conform to the expectancies held by O and vice versa, creating a stable interaction pattern. When O's behavior falls outside the range of expected behavior, it is classified as a violation. Communication expectations are not exact; instead, an individual has a range or tolerance level for a communication expectation. Figure 2.4 illustrates this range. The violations are valenced as positive or negative. How they are valenced partly depends on who is committing the violation. If the violator is someone who is regarded favorably—with what is called high communicator reward valence—the violation may also be regarded as a positive act. If the violator is someone held in low regard, the violation may be regarded as a negative one. For example, if a positively regarded O moves very close to S, this invasion of personal space may be interpreted by S as a show of affection or affiliation, making it a positive violation. If a negatively regarded O commits the same personal space invasion, it may be interpreted as a threat, making it a negative violation.

The valencing of violations is actually preceded by the degree to which the violation triggers arousal and an appraisal process. Violations are thought to be *arousing* and *uncertainty-provoking*, resulting in heightened attention to the meaning of the violation and its desirability. This is the basis for the first propositions of the spiral model of trust, derived from EVT, that are testable:

P1: Behavioral violations of expectations elicit increased arousal compared to behavioral expectancy confirmation.

Fig. 2.4 Valencing of expectancy confirmations and violations

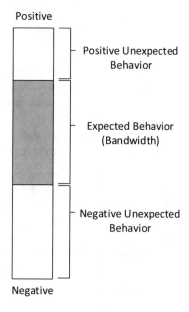

Continuum of Expectancies

Positive

Positive Unexpected Behavior

Expected Behavior (Bandwidth)

Negative Unexpected Behavior

Negative

P2: Violations of expectations elicit increased uncertainty compared to behavioral confirmation.

This arousal is thought to elicit a bipartite *appraisal* process such that S draws implicit interpretations of the violative act and evaluates it as welcome or not. The appraisal process leads to a valencing of the violation (or confirmation) as positive or negative. These appraisals can occur instantaneously and simultaneously, and as an overlearned process need not be cognitively burdensome. Different types of violations have consensually-understood meanings within a language community, and the likelihood of them being judged as desirable or not is also normative within a language community. For example, a male O putting an arm around a woman coworker S's shoulder can be variously interpreted as signaling condescension, congratulations, or flirtation. Whether it is welcome or not may be dictated by whether S views O as someone of higher status or someone held in high regard. A condescending touch will be unwelcome, a congratulatory touch will be welcome, and a flirtatious touch will be only welcome from a highly regarded O.

P3: Verbal and nonverbal violations have unique and identifiable meanings associated with them.
P4: Violations can be evaluated on a continuum from not at all welcome to completely welcome.

A novel aspect of EVT is that it introduces the concept of positive violations and distinguishes positive and negative violations from confirmations that have a positive or negative valence. Confirmations are behaviors that fall within the expected range. EVT predicts that positive violations produce more desirable results than positive confirmations, so trust should be highest when a positive violation has occurred. Conversely, trust should be lowest when a negative violation has occurred (although a negative confirmation could also occur). In this regard, trust is an integral concern that emerges as a consequence of expectancies being met or disconfirmed:

P5: Trust is positively related to positive expectancy violations and inversely related to negative expectancy violations.
P6: Positive verbal and nonverbal violations by O engender more trust by S than positive confirmations.
P7: Negative verbal and nonverbal violations by O engender more suspicion by S than negative confirmations.

One caveat to negative violations is that first instances with a highly regarded O may be uncertainty-provoking and require multiple instances before being registered as a negative violation (Afifi and Burgoon 2000). Put differently, negative violations are not as predictable as positive violations. Single instances of violations may produce a range of reactions, the impact of which depends on the reward value of the perpetrator of the violation. Single negative violations may be disregarded, misinterpreted, or misattributed. The same is not true if the perpetrator is poorly regarded. In that case, a violation may have an amplified effect. In other words, the perpetrator's reward value multiplies the effect of a violation. We revisit this issue shortly, as it relates to the concept of a spiral.

Early explications of EVT (e.g., Burgoon 1983, 1993; Burgoon and Burgoon 2001; Burgoon and Le Poire 1993; Burgoon and Walther 1990; Le Poire and Burgoon 1994) drew attention to normative and individuated expectations, and the consequences associated with those violations. But those expositions did not address the interaction process that emerges. Burgoon et al. (1995a) extended EVT to the communication process by predicting the circumstances under which a violation by O would cause S to reciprocate or compensate for O's level of nonverbal involvement and pleasantness. Burgoon et al. (2016) attempted to further elaborate on the constituent elements of the communication process by incorporating three key nonverbal aspects: perceived involvement, perceived mutuality, and ease of coordination between S and O. Conversational involvement refers to the degree to which participants in a communicative exchange are cognitively and behaviorally engaged in the topic, relationship and/or situation (Coker and Burgoon 1987). Perceived mutuality encompasses feelings of connectedness, receptivity, and mutual understanding that contribute to a sense of "relationship" or "groupness" among participants (Burgoon et al. 2010). Ease of coordination refers to the ease, naturalness and fluidity of the interaction process (Burgoon et al. 2002). These cognitive-emotive judgments of O made by S in turn affect task and social consequences of the interaction, one of which is trust. We predict that these three perceptions related to the interaction process should all promote trust because they reinforce a sense of connectedness and common ground.

P8: Trust is positively related, and distrust inversely related, to perceived conversational involvement.

P9: Trust is positively related, and distrust inversely related, to perceived mutuality.

P10: Trust is positively related, and distrust inversely related, to ease of coordination.

Additionally, repeated experience of a positive or negative violation should contribute to a spiral of trust growing, in the case of positive actions by the perpetrator, and erosion of trust by repeated negative actions. Although both positive confirmations and positive violations should have beneficial effects on trust, positive violations should have a greater impact on trust because of the arousal and appraisal processes described above. Whether the opposite is true for negative violations is less clear because confirming a negative expectation in itself should damage trust, and a spiral of such actions may quickly reach the floor for distrust.

Notwithstanding the maturing of EVT over the course of four decades, gaps remain. EVT is silent on how one's preferences and pressing needs affect the trust process. EVT also focuses more on nonverbal than verbal behaviors, and it does not handle moment-to-moment changes in interaction behavior. These factors were among the impetus for the development of *Interaction Adaptation Theory* (IAT; Burgoon et al. 1993; Burgoon et al. 1995a). IAT was also a response to the inadequacies of previous models (such as affiliative conflict theory and norm of reciprocity) to account for what factors lead communicators to adapt (or not) to one another's verbal and nonverbal behavior and with what consequences. For example, dyadic interaction models emphasize an overly simplistic, single causal mechanism and cannot account for the broad array of behavior patterns that are observed in routine discourse or repeated interactions.

IAT derived its principles from a synthesis of biological, psychological, socio-logical, and communicational models of human interpersonal interaction, with an eye toward producing not only greater definitional clarity but also testable predic-tions and explanations of the observed communication patterns in human inter-changes. Although its progenitors were models predicting patterns of reciprocal and compensatory dyadic interaction, IAT went a step farther in linking those patterns to outcomes, thus making it a strong candidate for predicting and explaining the devel-opment, maintenance, and erosion of trust.

The primary concepts in IAT are the actor variables of *requirements, expecta-tions*, and *desires*; an *interactional position (IP)*; and *actual performed behavior (AP)*. The three interrelated actor variables that IAT postulates influence interaction behavior significantly are requirements, expectations, and desires. *Requirements (R)* refer to biological drive states and deeply ingrained psychological needs such as safety, nourishment, and respite. Humans who are fearful, stressed, hungry, fatigued, and so on will be motivated to alleviate these needs above all else. If basic needs are not satisfied, their fulfillment will drive behavior, leading to instinctive fight, freeze, or flight responses. For instance, a captive fearing for his or her life and dependent on a captor for survival may say or do whatever is perceived to improve chances of survival and may, ironically, come to place trust in the captor (the Stockholm syndrome).

Some requirements are universal. For example, all humans seek protection from harm, and when physical security is at stake, determining others' trustworthiness may be of paramount concern. Other needs are linked to culture, socio-demographics, or personality. For example, people with collectivist cultural orientations may seek more identification and inclusion in their cultural group than people with individu-alistic orientations, whereas the latter may be more driven to seek autonomy and independence of action (Triandis 1972, 1994). Men need more personal space and less crowding than women. Introverts need more solitude than extroverts, and so on (Burgoon 1983). Some of these needs are static during the course of a single encoun-ter. Others are changeable. For example, fatigue may grow over a lengthy, boring meeting; fear may grow across the course of an interrogation.

The conceptualization of *expectations* (E) is taken from EVT. E may be norm-based or person-specific. Inasmuch as culture-linked and gender-linked expecta-tions are potent influences on interaction patterns, meeting these expectations can foster trust; deviating from them can prompt distrust. Members of collectivist cul-tures expect more indirect speech and less self-promotion than do members of indi-vidualist cultures (Gudykunst and Kim 1992). Examples of this phenomenon abound. As compared to Asian cultures, women in western cultures are expected to interact at closer distances than are men. Middle-eastern men who approach western men in close proximity may be distrusted, while simultaneously any distancing moves by western men may trigger suspicion that they are hiding something from their Middle-eastern interlocutors (Neuliep 2017). In hierarchically-oriented cultures with large power distances, high-status in-group members expect low-status outgroup members to show them respect and deference in interactions; failure to do so is also a basis for distrust (see, e.g., Beatty 2001; Gudykunst et al. 1996; Kupperbusch et al. 1999; Wolfgang 1984).

Norms, customs, and rules for a given communication context should be particularly relevant to setting expectations. Established protocols, self-presentation demands, requirements for conversation management for the type of episode in force, emotional regulation, and the like should be widely understood and adhered to. Whereas compliance with norms does not necessarily earn trust, negative violations of prevailing norms may create suspicion and distrust because, in part, norm violations identify the violator as an out-group member or someone whose behavior is unpredictable or unexpected. For example, the loud and expansive gesturing of an Arab roundtable guest compared to the quiet reserve of a Japanese one may deflate trust among Japanese audience members but inflate it among Arab ones. It may reinforce the sense that trusting people whose behavior is nonnormative is more risky than trusting people whose behavior is predictable.

Other expectations are person-specific and relate to the known typical interaction patterns of the individual. If Self (S) is familiar with Other (O) and has a prior history with O, S will hold individuated expectations of O. Lacking such familiarity or experience, S's expectations will devolve to the social norms for O's personal characteristics, the relationship between S and O, and the context. Thus, in routine social interaction, much of the variance in behavior should be predictable from expectations, and both group-wide and personalized expectations should be empirically verifiable.

Desires (D) refer to individual goals, motives, preferences, and such (although desires may be shaped partially by culture, socio-demographics, and personality). People who are motivated for self-gain, for example, will communicate differently than those who are motivated altruistically for others' benefit. Humans are assumed to be goal-oriented and motivated to behave in ways that maximize their chances of achieving their goals. For example, in cultures that value harmonious interaction, members will be motivated to adopt communication patterns that minimize face threats. Motivations are necessarily linked to incentives and anticipated consequences. In general, people are motivated to avoid aversive consequences and to seek beneficial consequences. In adversarial relationships between the military and local citizenry in an occupied country, for instance, the risks of retaliation against those who cooperate with military personnel must be juxtaposed against the potential jeopardy for failure to cooperate. The relative weighting of those costs and benefits will motivate the person's communication.

RED are hierarchical in their influence on interaction, with requirements taking precedence over expectations and desires, and expectations taking precedence over desires. *All propositions related to expectations and motivations assume that requirements have been met; if requirements are not met, they supersede other factors in governing behavior.*

Although not specified in the original explication of IAT, the jointly-defined *relationship* between S and O is a further, multifaceted class of characteristics related to the actors. One way to characterize the relationship is by standard sociodemographic and relational categories such as same- versus mixed-sex, friend or foe, stranger versus familiar, status-equal or -unequal, superior-subordinate, and so forth. Burgoon and Hale's (1984) relational topoi are another way to characterize relationships along continua of dominant-submissive, affectionate-hostile (liked versus disliked, cooperative versus antagonistic), deep-superficial (familiar versus unfamiliar), inclusive-exclusive, similar-dissimilar, composed-tense, and so forth.

One of the topoi that defines relationships is trust-distrust. Trust is one of the enduring characteristics that define interpersonal relationships; trust becomes both cause and effect as it cycles through a relationship. The nature of the relationship, especially its valence, power equality, and inclusiveness, is a significant driver of the initial interaction position, described next, and resultant behavior patterns by virtue of influencing what the respective actors need, expect, and desire.

Interaction Position, IP, is a concept from IAT that expresses the net combination of all the exogenous variables (*RED*). It originally described a person's starting place at the outset of any given interaction; i.e., it was meant to describe behavioral predispositions and concomitant physiology, psychological states, and cognitive states at the start of the interaction, known as Time-zero (T_0). However, the *IP* can also refer to the beginning point of any episode that is within an interaction comprised of multiple episodes, phases, or topics. Because the spiral model of trust emphasizes communication patterns, the model shows these internal states separately from the behavioral states, but they are assumed to accompany the changing communication landscape.

The *IP* is the net quotient of combining the various *RED* factors. For example, if S expects a cooperative social chat with another liked ingroup member, O, who is of higher status, S should enter the interaction with the intention to exhibit approach behaviors and a customary demeanor of respect. S should also be in a non-agitated physiological state, with minimal cognitive load, hold favorable attitudes toward O, and be truth biased. If S instead expects O to be an adversarial outgroup authority figure with ability to apply punishment, S may begin the interaction in an agitated, fearful state that is accompanied by freeze or flight rather than fight behavior. In these latter cases, the relationship already predisposes S to be distrustful, but the interaction patterns that transpire could still alter that dynamic.

P11: S's evaluation of O's IP at T_0 affects S's verbal and nonverbal behavior at T_1.

Because the *IP* must be viewed as the "on balance" quotient of all the preceding factors, it might seem difficult to compute computationally. However, oftentimes the most salient factors will be self-evident. Prior knowledge of the relationship's state may make clear that S and O have a trusting relationship as they begin any communication episode, and the absence of any observable indicators of stress should reinforce assumptions about the degree of trust at T_0. Conversely, if two people are engaged in a conflict, one might assume that the adversaries will enter interactions with a low degree of trust, and building trust may be one of the first objectives.

Actual Performance, AP, refers to what the actors actually say and do. It includes nonverbal behaviors, verbal content, the juxtapositions of nonverbal and verbal elements vis a vis each other (e.g., whether statements show consistency or inconsistencies), and the degree of behavioral coordination and adaptation between S and O. Behavioral coordination includes whether S and O show other patterns of reciprocity or compensation and whether their interaction is coordinated and synchronous or not. These patterns are the most proximal, and potentially potent, determinants of trust.

IAT posits that S's RED values will produce S's *IP*, which will be compared to O's *AP*. Whichever is more favorably valenced will dictate approach or avoidance behaviors. Suppose S requires, expects, and desires a formal interaction. Her *IP* will be one of formal demeanor. Suppose that O's *AP* is an informal one that includes very relaxed posture, familiar forms of address, profane language, and the like. IAT predicts that S will negatively evaluate O's *AP* relative to S's *IP*, leading S to engage in avoidant behavior, such as displeased facial and vocal expressions, large conversational distance, hyper-formal language, and moves to end the conversation altogether. If O, by contrast, negatively evaluates S's *AP* relative to O's *IP*, he may respond by becoming even more informal to try to model the behavior he desires from S in the hopes of bringing their behavior patterns into alignment. Both are likely to come away from such a poorly coordinated interaction with distrust and dislike for the other.

P12: If the IP for O is favorable at T_0, S's nonverbal behavior will be to approach O's AP at T_1.

P13: If the IP for O is negative at T_0, S's nonverbal behavior will be avoidance of O's AP at T_1.

P14: S's approach toward/avoidance of O at T_1 will be correlated with trust/distrust for O at T_2.

P15: Repeated approach strengthens trust; repeated avoidance weakens trust.

Put in EVT terms, when the cycles of interaction exhibited by O repeatedly violate S's expectations negatively, this should produce negative outcomes such as distrust and dislike, as shown in Fig. 2.5. The same is true for O. If O's expectations are also violated negatively, this may lead not only to an asynchronous interaction pattern but also distrust of S's disengagement and stand-offish style.

If over the course of many conversational turns S and O are unable to adapt to one another's interaction styles, their failure to achieve interactional coordination and synchrony may become a source of interpersonal distrust. This is the spiral of trust and distrust that develops over multiple interaction episodes.

Suspicion refers to a state of uncertainty about another's character and behavioral intentions. It can spring forth from a variety of pre-interactional (e.g., interaction position, previous trust level) and interactional (e.g., verbal message content, ancillary nonverbal behavior) sources. Suspicion's relationship to trust is curvilinear in that it is associated with the highest degree of uncertainty (Burgoon et al. 1996). As

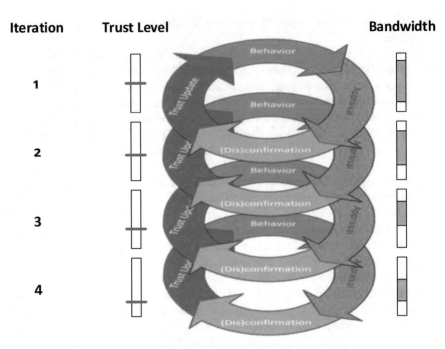

Fig. 2.5 Downward iterations of distrust formation based on behavior, attribution, and expectancy confirmations and disconfirmations

Fig. 2.6 Increasing
scrutiny during appraisal

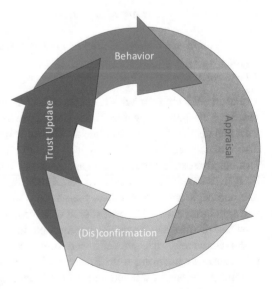

uncertainty is reduced, suspicion either morphs into distrust (greater certainty about the other's untrustworthiness) or trust (greater certainty about the other's trustworthiness). To reduce uncertainty, S may invest more cognitive effort to size up O's behavior during the appraisal stage, as shown in Fig. 2.6. With additional scrutiny (and perhaps further information-gathering), S may alleviate the uncertainty and transform suspicion into distrust or trust.

Within an interaction, suspicion is most likely to be low when O and S are synchronized in their exchanges. Burgoon et al. (2017) describe how interactants achieve coordination and adaptation in their interactions. Interactional synchrony is achieved when S and O use language and exhibit behavior that converges toward, rather than diverges away from, one another. When both S and O adopt a converging interaction pattern, they are exhibiting an approachable stance and signaling to each other that they desire to be trusted. In other words, through interactional synchrony, S (or O) may send trusting overtures that are then returned to O (or S) by way of behavioral adaptation and accommodation. In contrast, diverging interaction patterns are likely to engender uncertainty and lead to suspicion (Dunbar et al. 2014).

How REDs between S and O relate to one another follows the hierarchical priorities. If Rs are active, they will dictate response patterns. If group members are hungry, for example, and O chooses to talk at length, S may compensate by making very short responses, cutting off turns at talk for O, and giving negative or dismissive nonverbal feedback. Under this kind of interaction pattern, trust is unlikely to grow. If Rs are not in play, S and O may follow EVT predictions such that positive violations elicit more trust than positive confirmations, whereas negative violations erode trust. Finally, where behavioral patterns are not constrained by expectations and Ss are free to act upon their Ds, increases in coordination and interactional synchrony will fuel more trust. Deceivers may capitalize upon these communication patterns by attempting to mirror O's nonverbal demeanor and adopt their verbal communication style. However, engaging in what communication accommodation theory calls hyper-accommodation by overdoing convergence may reach a point of sycophancy and appearing inauthentic.

P16: *Interactional dissynchrony contributes to uncertainty and reduces the level of trust.*

P17: *The development of interactional synchrony through reciprocity and accommodation of positively valued communication patterns engenders trust.*

Garland et al. (2010) argue that self-perpetuating and damaging cycles triggered by negative emotions are like *downward spirals,* whereas self-perpetuating cycles that capitalize on positive emotions and lead to optimal functioning and enhanced social openness are referred to as *upward spirals* (see Fig. 2.7). Using the logic of EVT and IAT, we believe that trust development can also be best viewed as a spiral that expands and contracts based on a number of factors. For example, if S has an existing positive and trusting relationship with O, then she will not need a lengthy appraisal process and will likely evaluate many of O's deviations from expectations favorably, leading to a reinforcement of trust. If S has reason to be suspicious of O based on previous interactions, she may scrutinize O's behavior more closely,

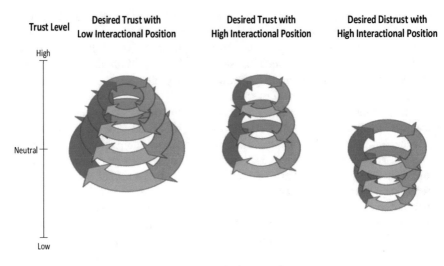

Fig. 2.7 Examples of upward and downward spirals

evaluate O's behavior deliberately, and extend the length of the trust spiral. Trust assessments are constantly being updated over the course of an interaction and relationship, leading to a repetitive spiral that modifies trust over time.

P18: Trust spirals expand and contract based on actual behaviors, expectancies based on norms or previous interactions, and appraisals of the IP and AP.

One aspect of interactions that can affect trust-distrust is interactional coordination and synchrony. Coordinated, meshed interaction and synchrony are positively related to trust. For example, scholars have developed data mining tools that uncover hidden behavioral patterns which reveal the extent to which interaction is organized between two people or lacks interdependence. A program called Theme that uncovers hidden behavioral patterns reveals the extent to which interaction is organized between two people or lacks interdependence (Burgoon et al. 2015). In deceptive interactions, even though at the conscious level S and O may be unaware of the extent to which they are achieving interdependence, their nonverbal coordination may signify the growth of trust over time. There is, however, a point beyond which hyper-accommodation appears forced and inauthentic, which backfires. Put differently, interaction synchrony is a positive force as long as it is fluent, rhythmic, and nonconscious. Once it reaches an upper limit, it draws attention to itself and loses its naturalness. Deceptive interchanges that overstep this limit may betray their lack of verisimilitude.

P19: The relationship between interactional coordination and trust is nonlinear such that at its upper reaches, trust reverses into distrust.
P20: Increases in trust reinforce perceived truthfulness.

Summary

Trust is an integral part of interpersonal relationships. Achieving trust is typically an assumption and a goal in most relationships, although there are occasions when wariness and distrust are instead the goal. Problematic are the cases when trust is desired but distrust prevails or distrust is wanted and undue trust instead transpires. These cases of how achieving or failing to achieve end desired states of trust or distrust are the objective of our integration of expectancy violations theory and interaction adaptation theory into a spiral model of trust. These theories seem most apropos because they recognize that humans are goal-oriented, they evaluate the actions of others, their communication is multimodal in the sense that there are multiple nonverbal and verbal signals from which to fashion messages, communication patterns may conform to or violate expectations, and patterns evolve over the course of single episodes or multiple episodes.

In relationships, expectations are founded on social norms and individuated projections for another's behavior based on prior experience and personalized knowledge. Repeated confirmations of positive expectations should build trust, whereas positive expectancy violations should speed up the trust-building process. Conversely, negative violations should erode trust.

The likelihood of S and O interaction patterns becoming coordinated, rhythmic, and interdependent will be governed by each person's requirements, expectations, and desires. These are synthesized into an interaction position—a projected verbal and nonverbal behavior pattern against which the other's actual behavior pattern will be compared. When the actual behavior pattern is more favorable than the projected pattern, the person will engage in approach behaviors, which typically elicit similar behaviors and a sense of trust. When the actual behavior pattern is more negative than the projected pattern, the person will engage in avoidance behaviors that are accompanied by distrust. Repeated iterations of these of these patterns become a positive or negative spiral.

The propositions advanced here are not meant to be a comprehensive enumeration of the propositions of the theory but rather the beginnings of it. We welcome additions and modifications by others to create a more robust spiral theory of trust. Tests of this model's propositions should enlighten the extent to which trust, once established, remains fairly fixed, or spirals over time in response to the verbal and nonverbal behaviors of participants. If trust fluctuates over time, the timing of its measurement becomes critical. Like taking a child's temperature who has the flu, periodic readings are required. In the context of the SCAN project, observation of patterns of dominance and arousal by participants over time may reveal whether those patterns are significant predictors of trust and are linked to deception.

Acknowledgement   We are grateful to the Army Research Office for funding much of the work reported in this book under Grant W911NF-16-1-0342.

Funding Disclosure   This research was sponsored by the Army Research Office and was accomplished under Grant Number W911NF-16-1-0342. The views and conclusions contained in this document are those of the authors and should not be interpreted as representing the official policies, either expressed or implied, of the Army Research Office or the U.S. Government. The U.S. Government is authorized to reproduce and distribute reprints for Government purposes notwithstanding any copyright notation herein.

References

Adler, T. R. (2007). Swift trust and distrust in strategic partnering relationships: Key considerations of team-based designs. *Journal of Business Strategies, 24*(2), 105–121.

Afifi, W. A., & Burgoon, J. K. (2000). The impact of violations on uncertainty and consequences for attractiveness. *Human Communication Research, 26,* 203–233.

Bambacas, M., & Patrickson, M. (2008). Interpersonal communication skills that enhance organisational commitment. *Journal of Communication Management, 12*(1), 51–72.

Beatty, J. (2001). Language and communication. In L. L. Adler & U. P. Gielen (Eds.), *Crosscultural topics in psychology* (2nd ed., pp. 47–59). Westport: Praeger Publishers/Greenwood Publishing Group.

Bernieri, F. J. (1988). Coordinated movement and rapport in teacher-student interactions. *Journal of Nonverbal Behavior, 12*(2), 120–138.

Buchan, N. R., Johnson, E. J., & Croson, R. T. (2006). Let's get personal: An international examination of the influence of communication, culture and social distance on other regarding preferences. *Journal of Economic Behavior & Organization, 60*(3), 373–398.

Burgoon, J. K. (1976). The ideal source: A reexamination of source credibility measurement. *Central States Speech Journal, 27,* 200–206.

Burgoon, J. K. (1983). Nonverbal violations of expectations. In J. Wiemann & R. Harrison (Eds.), *Nonverbal interaction: Vol. 11. Sage annual reviews of communication* (pp. 11–77). Beverly Hills: Sage.

Burgoon, J. K. (1993). Interpersonal expectations, expectancy violations, and emotional communication. *Journal of Language and Social Psychology, 12,* 30–48.

Burgoon, J. K., & Burgoon, M. (2001). Expectancy theories. In P. Robinson & H. Giles (Eds.), *Handbook of language and social psychology* (2nd ed., pp. 79–101). Sussex: Wiley & Sons.

Burgoon, J. K., & Hale, J. L. (1984). The fundamental topoi of relational communication. *Communication Monographs, 51,* 193–214.

Burgoon, J. K., & Hale, J. L. (1988). Nonverbal expectancy violations: Model elaboration and application to immediacy behaviors. *Communications Monographs, 55,* 58–79.

Burgoon, J. K., & Le Poire, B. A. (1993). Effects of communication expectancies, actual communication, and expectancy disconfirmation on evaluations of communicators and their communication behavior. *Human Communication Research, 20,* 75–107.

Burgoon, J. K., & Walther, J. B. (1990). Nonverbal expectancies and the consequences of violations. *Human Communication Research, 17,* 232–265.

Burgoon, J. K., Dillman, L., & Stern, L. A. (1993). Adaptation in dyadic interaction: Defining and operationalizing patterns of reciprocity and compensation. *Communication Theory, 3,* 196–215.

Burgoon, J. K., Le Poire, B. A., & Rosenthal, R. (1995a). Effects of preinteraction expectancies and target communication on perceiver reciprocity and compensation in dyadic interaction. *Journal of Experimental Social Psychology, 31,* 287–321.

Burgoon, J. K., Stern, L. A., & Dillman, L. (1995b). *Interpersonal adaptation: Dyadic interaction patterns.* Cambridge, UK: Cambridge University Press.

Burgoon, J. K., Buller, D. B., Ebesu, A., Rockwell, P., & White, C. (1996). Testing interpersonal deception theory: Effects of suspicion on nonverbal behavior and relational messages. *Communication Theory, 6,* 243–267.

Burgoon, J. K., Bonito, J. A., Bengtsson, B., Ramirez, A., Jr., Dunbar, N. E., & Miczo, N. (1999). Testing the interactivity model: Communication processes, partner assessments, and the quality of collaborative work. *Journal of Management Information Systems, 16*(3), 33–56.

Burgoon, J. K., Bonito, J. A., Ramirez, A., Kam, K., Dunbar, N., & Fischer, J. (2002). Testing the interactivity principle: Effects of mediation, propinquity, and verbal and nonverbal modalities in interpersonal interaction. *Journal of Communication, 52,* 657–677.

Burgoon, J. K., Chen, F., & Twitchell, D. (2010). Deception and its detection under synchronous and asynchronous computer-mediated communication. *Group Decision and Negotiation, 19,* 346–366.

Burgoon, J. K., Wilson, D., Hass, M., & Schuetzler, R. (2015). Interactive deception in group decision-making: New insights from communication pattern analysis. In M. Magnusson, J. K. Burgoon, & M. Casarrubea (Eds.), *Discovering hidden temporal patterns in behavior and interaction: T-pattern detection and analysis with THEME.* New York: Springer.

Burgoon, J. K., Bonito, J. A., Lowry, P. B., Humpherys, S. L., Moody, G. D., Gaskin, J. E., & Giboney, J. S. (2016). Application of expectancy violations theory to communication with and judgments about embodied agents during a decision-making task. *International Journal of Human-Computer Studies, 91,* 24–36.

Burgoon, J. K., Dunbar, N. E., & Giles, H. (2017). Interaction coordination and adaptation. In A. Vinciarelli, M. Pantic, N. Magnenat-Thalmann, & J. K. Burgoon (Eds.), *Social signal processing* (pp. 78–96). Cambridge, UK: Cambridge University Press.

Coker, D. A., & Burgoon, J. K. (1987). The nature of conversational involvement and nonverbal encoding patterns. *Human Communication Research, 13,* 463–494.

Cook, K. S., Levi, M., & Harden, R. (Eds.). (2009). *Whom can we trust? How groups, networks, and institutions make trust possible.* New York: Russel Sage Foundation.

Duggan, A. P., & Parrott, R. L. (2001). Physicians' nonverbal rapport building and patients' talk about the subjective component of illness. *Human Communication Research, 27*(2), 299–311.

Dunbar, N. E., Jensen, M. L., Tower, D. C., & Burgoon, J. K. (2014). Synchronization of nonverbal behaviors in detecting mediated and non-mediated deception. *Journal of Nonverbal Behavior, 38*(3), 355–376.

Fiksdal, S. (1988). Verbal and nonverbal strategies of rapport in cross-cultural interviews. *Linguistics and Education, 1*(1), 3–17.

Foddy, M., Platow, M. J., & Yamagishi, T. (2009). Group-based trust in strangers: The role of stereotypes and expectations. *Psychological Science, 20*(4), 419–422.

Garland, E. L., Fredrickson, B., Kring, A. M., Johnson, D. P., Meyer, P. S., & Penn, D. L. (2010). Upward spirals of positive emotions counter downward spirals of negativity: Insights from the broaden-and-build theory and affective neuroscience on the treatment of emotion dysfunctions and deficits in psychopathology. *Clinical Psychology Review, 30*(7), 849–864.

Gilbert, D. T., Krull, D. S., & Malone, P. S. (1990). Unbelieving the unbelievable: Some problems in the rejection of false information. *Journal of Personality and Social Psychology, 59*(4), 601–613.

Gouldner, A. W. (1960). The norm of reciprocity: A preliminary statement. *American Sociological Review,* 161–178.

Grice, P. (1989). *Studies in the way of words.* Cambridge, MA: Harvard University Press.

Gudykunst, W. B., & Kim, Y. Y. (1992). *Communicating with strangers: An approach to intercultural communication* (2nd ed.). New York: McGraw Hill.

Gudykunst, W., Matsumoto, Y., Ting-Toomey, S., Nishida, T., Kim, K., & Heyman, S. (1996). The influence of cultural individualism-collectivism, self-construals and individual values on communication styles across cultures. *Human Communication Research, 22,* 510–543.

Heintzman, M., Leathers, D. G., Parrott, R. L., & Cairns, A. B., III. (1993). Nonverbal rapport-building behaviors' effects on perceptions of a supervisor. *Management Communication Quarterly, 7*(2), 181–208.

Hinsz, V. B., Tindale, R. S., & Vollrath, D. A. (1997). The emerging conceptualization of groups as information processors. *Psychological Bulletin, 121*(1), 43–64.

Ho, T. H., & Weigelt, K. (2005). Trust building among strangers. *Management Science, 51*(4), 519–530.

Hovland, C. I., Janis, I. L., & Kelley, H. H. (1953). *Communication and persuasion*. New Haven: Yale University Press.

Iacono, C. S., & Weisband, S. (1997). Developing trust in virtual teams. In *Proceedings of the thirtieth Hawaii international conference on system sciences* (Vol. 2, pp. 412–420). Los Alamitos: IEEE.

Jarvenpaa, S. L., Knoll, K., & Leidner, D. E. (1998). Is anybody out there? Antecedents of trust in global virtual teams. *Journal of Management Information Systems, 14*(4), 29–64.

Jung, D. I., & Avolio, B. J. (2000). Opening the black box: An experimental investigation of the mediating effects of trust and value congruence on transformational and transactional leadership. *Journal of Organizational Behavior, 21*(8), 949–964.

Kim, J. S., Weisberg, Y. J., Simpson, J. A., Oriña, M. M., Farrell, A. K., & Johnson, W. F. (2015). Ruining it for both of us: The disruptive role of low-trust partners on conflict resolution in romantic relationships. *Social Cognition, 33*(5), 520–542.

Koeszegi, S. T. (2004). Trust-building strategies in inter-organizational negotiations. *Journal of Managerial Psychology, 19*(6), 640–660.

Kramer, R. M., Brewer, M. B., & Hanna, B. A. (1996). Collective trust and collective action: The decision to trust as a social decision. In R. M. Kramer & T. R. Tyler (Eds.), *Trust in organizations: Frontiers of theory and research* (pp. 357–389). Thousand Oaks: Sage.

Kupperbusch, C., Matsumoto, D., & Kooken, K. (1999). Cultural influences on nonverbal expressions of emotion. In P. Philippot, R. S. Feldman, & E. J. Coats (Eds.), *The social context of nonverbal behavior* (pp. 17–44). New York: Cambridge University Press.

Le Poire, B. A., & Burgoon, J. K. (1994). Two contrasting explanations of involvement violations: Expectancy violations theory versus discrepancy arousal theory. *Human Communication Research, 20*, 560–591.

Lewicki, R. J., & Bunker, B. B. (1995). Trust in relationships: A model of development and decline. In B. B. Bunker & J. Z. Rubin (Eds.), *The Jossey-Bass conflict resolution series. Conflict, cooperation, and justice: Essays inspired by the work of Morton Deutsch* (pp. 133–173). San Francisco: Jossey-Bass.

Lewicki, R. J., McAllister, D. J., & Bies, R. J. (1998). Trust and distrust: New relationships and realities. *Academy of Management Review, 23*(3), 438–458.

Manusov, V. (1992). Mimicry or synchrony: The effects of intentionality attributions for nonverbal mirroring behavior. *Communication Quarterly, 40*(1), 69–83.

Manusov, V., Winchatz, M. R., & Manning, L. M. (1997). Acting out our minds: Incorporating behavior into models of stereotype-based expectancies for cross-cultural interactions. *Communications Monographs, 64*(2), 119–139.

Mayer, R. C., Davis, J. H., & Schoorman, F. D. (1995). An integrative model of organizational trust. *Academy of Management Review, 20*(3), 709–734.

McCroskey, J. C., & Young, T. J. (1981). Ethos and credibility: The construct and its measurement after three decades. *Communication Studies, 32*(1), 24–34.

Meyerson, D., Weick, K. E., & Kramer, R. M. (1996). Swift trust and temporary groups. In R. M. Kramer & T. R. Tyler (Eds.), *Trust in organizations: Frontiers of theory and research* (pp. 166–195). Thousand Oaks: Sage.

Neuliep, J. W. (2017). *Intercultural communication: A contextual approach*. Los Angeles: Sage Publications.

Peters, K., & Kashima, Y. (2007). From social talk to social action: Shaping the social triad with emotion sharing. *Journal of Personality and Social Psychology, 93*(5), 780–797.

Rempel, J. K., Ross, M., & Holmes, J. G. (2001). Trust and communicated attributions in close relationships. *Journal of Personality and Social Psychology, 81*(1), 57–64.

Robert, L. P., Denis, A. R., & Hung, Y. T. C. (2009). Individual swift trust and knowledge-based trust in face-to-face and virtual team members. *Journal of Management Information Systems, 26*(2), 241–279.

Rousseau, D. M., Sitkin, S. B., Burt, R. S., & Camerer, C. (1998). Not so different after all: A cross-discipline view of trust. *Academy of Management Review, 23*(3), 393–404.

Sagarin, B. J., Rhoads, K. V. L., & Cialdini, R. B. (1998). Deceiver's distrust: Denigration as a consequence of undiscovered deception. *Personality and Social Psychology Bulletin, 24*(11), 1167–1176.

Silvester, J., Patterson, F., Koczwara, A., & Ferguson, E. (2007). "Trust me...": psychological and behavioral predictors of perceived physician empathy. *Journal of Applied Psychology, 92*(2), 519–527.

Simpson, J. A. (2007). Psychological foundations of trust. *Current Directions in Psychological Science, 16*(5), 264–268.

Taylor, D. A., & Altman, I. (1987). Communication in interpersonal relationships: Social penetration processes. In M. E. Roloff & G. R. Miller (Eds.), *Sage annual reviews of communication research, Vol. 14. Interpersonal processes: New directions in communication research* (pp. 257–277). Thousand Oaks: Sage Publications.

Teven, J. J., & McCroskey, J. C. (1997). The relationship of perceived teacher caring with student learning and teacher evaluation. *Communication Education, 46*(1), 1–9.

Tickle-Degnen, L., & Rosenthal, R. (1990). The nature of rapport and its nonverbal correlates. *Psychological Inquiry, 1*(4), 285–293.

Triandis, H. C. (1972). *The analysis of subjective culture.* New York: McGraw-Hill.

Triandis, H. C. (1994). *Culture and social behavior.* New York: McGraw-Hill.

Van Overwalle, F., & Heylighen, F. (2006). Talking nets: A multiagent connectionist approach to communication and trust between individuals. *Psychological Review, 113*(3), 606–627.

Wolfgang, A. (Ed.). (1984). *Nonverbal behavior: Perspectives, applications, intercultural insights.* Lewiston: C.J. Hogrefe.

Yuki, M., Maddux, W. W., Brewer, M. B., & Takemura, K. (2005). Cross-cultural differences in relationship-and group-based trust. *Personality and Social Psychology Bulletin, 31*(1), 48–62.

Chapter 3
The Impact of Culture in Deception and Deception Detection

Matt Giles, Mohemmad Hansia, Miriam Metzger, and Norah E. Dunbar

In a rapidly globalizing world, communication across cultures is increasingly more common. Given the inevitability of miscommunication or deception in some cross-cultural communication (Levine et al. 2016), it is beneficial to understand the culturally based norms, values, and communicative styles of conversational partners (Holliday 2016). However, the exact relationship between culture and deception is complex. Further, applying existing research on deception detection and culture is a fraught task due to the differences between cross-cultural deception and cultural ingroup deception strategies. Put more simply, people lie to different people for different reasons and in different ways (Taylor et al. 2017). As such, deception cannot be treated as though it happens uniformly both across and within cultures. Due to the high stakes inherent to deception detection faced by members of the military and others in international contexts where intercultural communication takes place, Yager et al. (2009) explain that misunderstandings, errors, and ignorance "can have disastrous consequences" (p. 1). And even in everyday life, understanding and being able to detect deception is both difficult and important in intercultural interactions. Recognizing the role of culture, the work in this volume takes up the call of Taylor et al. (2017) to study deception in a wide array of contexts to uncover important yet undiscovered cultural effects.

The Socio-cultural Attitudinal Network (SCAN) project described in this volume was conceived to fill a gap in our knowledge because most deception research has been done in a "cultural vacuum" (Castillo 2015). The vast majority of studies on verbal and nonverbal cues to deception or deception detection skill have been done in English-speaking, Western cultures. Very few studies examine *cross-cultural* differences in displays associated with deception or the detection of deception (i.e., comparing norms and behaviors of people who are situated in different cultures,

M. Giles · M. Hansia · M. Metzger (✉) · N. E. Dunbar
University of California, Santa Barbara, Santa Barbara, CA, USA
e-mail: metzger@ucsb.edu

© Springer Nature Switzerland AG 2021
V. S. Subrahmanian et al. (eds.), *Detecting Trust and Deception in Group Interaction*, Terrorism, Security, and Computation,
https://doi.org/10.1007/978-3-030-54383-9_3

such as cues used during deception by people in China versus people in Spain), and even fewer have examined *intercultural* interactions in which members from different cultures interact (e.g., when a person from China interacts with a person from Spain). Some studies argue that deception has vast similarities across cultures, such as in the Global Deception Research Team's (2006) study that revealed the persistence of myths about eye gaze and other unreliable cues across cultures. There are two main perspectives on how detection of deception works across cultural contexts. One perspective, known as the specific discrimination perspective (Bond et al. 1990; Castillo 2015), indicates that differences in language and culture make deception detection difficult. On the other hand, the universal cue perspective suggests the possibility of a collection of indicators of deception which would be true across cultures (Al-Simadi 2000; Bond and Atoum 2000). Both these similarities and the accompanying emergent differences will be addressed in this chapter, dealing with the current state of deception detection scholarship within and across cultural contexts.

This chapter begins with a discussion of how culture is defined and the way in which this definition determines the context for what is and what is not considered to be deception. Next, it explores how culture has been studied previously in the context of deception and addresses variations in those analyses. After establishing the current state of research on deception detection by reviewing recent advances, this chapter next explores the role of cultural differences and similarities in modern theorizing about deception. Finally, we note the challenges inherent to bridging intercultural communication studies and deception detection research, outline both the pitfalls and best practices for this work, and illustrate how we have chosen to conduct our research.

How Have Researchers Operationalized Culture when Studying Deception?

Culture is a learned meaning system that consists of patterns of traditions, beliefs, values, norms, meanings, and symbols that are passed on from one generation to the next and are shared to varying degrees by interacting members of a community (Ting-Toomey et al. 2000). Although culture is viewed as a fairly stable characteristic of individuals and groups, Matsumoto et al. (1996) demonstrate that culture can be somewhat fluid with age, which reflects the ability of individuals to assimilate aspects of their nonnative cultural residences. Culture influences not only how verbal and nonverbal messages are produced, but also how they are perceived and interpreted, and with what consequences (Krys et al. 2016). Culture is immersive and cannot be entirely understood when disassociated from its proper context. McLuhan (1970) quips an old saying, "we don't know who discovered water, but we are sure it wasn't a fish!" (p. 2). This illustrates the way that culture surrounds us and provides the context through which all else is understood.

Given this complexity, operationalizing culture is a difficult task, which has been done in a variety of ways by different researchers. We will discuss below first how the relationship between individuals changes the nature of an interaction, as well as what can be extrapolated out from that interaction, and then identify different ways in which cultural designations are made.

First, it is important to consider the relationship between interlocuters in an interaction. Their ingroup/outgroup relationship is going to determine whether their interaction involves either (1) *intra*cultural deception, wherein individuals from *within the same cultural group* engage in deception within their own context; (2) or *inter*cultural deception, which involves individuals *from different cultural groups* engaging in deception within the same interaction. Deception can also be examined (3) *cross*-culturally when deceptive communication within a given culture is compared to another culture but members of each culture do not interact with one another. Because people behave differently in these different contexts, it cannot be assumed that the way in which an individual engages in deception in each context is going to be the same (Bradac et al. 1986). Demonstrating this, Whitty and Carville (2008) found that people lie differently with outgroup members than with ingroup members, specifically in that they use self-serving lies more with outgroup members. Individuals tell fewer lies to ingroup members and feel more uncomfortable when lying to them, as opposed to outgroup members (DePaulo and Kashy 1998).

Much of the research frames the way that individuals engage in deception differently in intercultural contexts, with the primary focus centering around managing anxiety and uncertainty (Gudykunst 2005). Broadly speaking, this perspective posits that because intercultural contexts involve engaging with others who operate from a different worldview, individuals may feel uncertain about how to interpret messages from people outside their culture and then how to meet the cultural expectations of others during the interaction—which in turn causes anxiety. In the process of managing this anxiety, participants interact with outgroup members differently than they do with ingroup members (Giles 2016). Notably, acting differently in an intercultural context is not necessarily a conscious decision. Sarbaugh (1979) argues that the degree of heterogeneity between two groups determines the level of 'interculturalness,' and that individuals subconsciously analyze and respond to this heterogeneity (Newmark and Asante 1976). Respectfully and fluidly responding to these differences is a key component of intercultural communication competence (Hammer et al. 1978).

Within intercultural interaction, researchers must decide how to classify each individual into specific cultural designations using one of multiple distinct – and often non-orthogonal – criteria. In order to classify individuals into different cultural groups, there are two main strategies that are relevant for intercultural deception in this context. These include (1) classifying individuals based on nationality or ethnic group, or (2) measuring culture at the individual level according to psychometric cultural dimension scales. Each of these operationalizations carries with it a specific set of assumptions that affect how the data can be applied to study culture. These two operationalizations are discussed below.

However, before doing so it is important to note that the aforementioned dynam-ics (i.e., the inter/intra-cultural relationship and the choice of cultural operational-ization) are not the only variables that shape the analysis of intercultural communication, but each combination of these variables has the possibility for unique constitutive rules which guide interaction that may not apply outside of their respective context. This makes cross-cultural comparisons challenging and makes it difficult to apply research findings that do not fit the same paradigm (Bond et al. 1990). Likewise, individuals demonstrate different motivations for lying in intra-group versus inter-group contexts (Dunbar et al. 2016).

Assigning cultural labels based on national identity In studies of deception in which there is a cultural component, culture is most often operationalized in terms of nationality. For example, Bond et al. (1990) studied how American versus Jordanian students lie about a person they liked and a person they disliked. Castillo (2011) compared Colombian and Australian liars. Lee et al. (1997) compared atti-tudes toward lying by Canadian and Chinese children. Leal et al. (2018) recently compared British, Chinese, and Arab liars. A similar approach can also be used to compare different subcultural groups within the same national culture. In such a study, Vrij and Winkel (1991) examined behavioral differences between white and black citizens in the South American country of Suriname. Commonly, these studies only compare two or possibly three countries at a time, and often the countries are selected due to convenience for the researchers rather than for testing theory. Castillo (2015) reviews these studies, finding few overall patterns in deception stud-ies that make between-country comparisons.

Measuring individual cultural dimensions as a cultural classification A sec-ond approach found in the deception literature is to examine what is termed "cul-tural dimensions." These include differences in demeanor (the way that people communicate with others) rooted in their cultural experiences. For example, Hall (1976) distinguished between "high-context" cultures and "low-context" cultures, which are defined by how explicitly and directly the people within these groups exchange information. People from a low-context culture will be more direct and verbal when conveying information because little clarifying information is avail-able in the context itself, whereas in a high-context culture, many things are left unsaid and it is up to the receiver to infer the intended meaning from nonverbal and contextual cues (Leal et al. 2018).

Hofstede (1980) differentiated cultures along a series of dimensions such as individualism-collectivism, high- and low-uncertainty avoidance, masculinity-femininity, and high- and low-power distance. His approach is widely used in psy-chology, sociology, marketing, communication, and management studies (Soares et al. 2007). Hofstede's original constructs have been refined to better reflect the subtleties of intercultural communication by social psychologists such as Triandis and Gelfland (Singelis et al. 1995; see Matsumoto et al. 1996, elaborating on using these dimensions to understand how culture impacts human interaction). Cultural dimensions can help to understand differences in the way that deception is both understood and interpreted in different cultures.

Of Hoftstede's dimensions, individualism-collectivism is probably the most studied in the context of deception. It relates to the degree to which individuals emphasize the needs and goals of the group over the needs and desires of individuals and relates to the degree of interconnectedness between group members. Collectivists are integrated into strong cohesive groups more than individualists (George et al. 2018). Kim et al. (2008) found that because collectivists value group harmony over individual needs, altering information in order to maintain group harmony is not always considered to be deceptive in collectivist societies. Thus, collectivists may experience less guilt or fear when lying than would individualists because it may be more acceptable to do so according to their cultural norms (Castillo 2015).

Other research finds that the cultural dimension of individualism/collectivism is related to trust in a way that could impact deception detection. Lowry et al. (2010) argue that collectivists' greater interdependence, which stems from valuing group rather than individual goals, and tighter social networks lead to a mindset that favors the development of interpersonal trust. Compared to individualists, collectivists place greater weight on social norms and opinions in judging the trustworthiness of others. This in turn facilitates trust transference between group members. Based on this logic, Lowry and colleagues hypothesized, and their data confirmed, that interpersonal trust was higher in collectivistic than in individualistic groups. George et al. (2018) similarly hypothesized that collectivists have a stronger sense of loyalty, respect, and trust toward others than individualists, making them less suspicious and therefore less likely to pay attention to leaked deception cues while communicating with others. He argued further that individualists may be less trusting of others and more prone to suspicion, making them better detectors of deception. While George et al. did not find support for this hypothesis, their study used primarily individualistic judges, which leaves open the question of what role collectivism might play in deception detection.

Another cultural dimension that relates to deception is the concept of "face." Ting-Toomey's (1988) face-negotiation theory emphasizes three face concerns during the resolution of interpersonal conflicts. Self-face is the concern for one's own image, for receiving approbation and for putting forth an impression of self that is socially favorable. Other-face is the concern for protecting another's image and protecting that self-presentation from threat. Mutual-face is concern for both parties' images and/or protecting the "image" of the relationship. The concept of face becomes especially problematic in situations with high uncertainty (such as embarrassment and conflict situations) when the situated identities of the communicators are called into question (Oetzel and Ting-Toomey 2003). Deception is one such situation because the parties are negotiating issues of trust and dependence in their interaction, and it might be highly face-threatening if one party does not believe the other. Oetzel and Ting-Toomey relate face to other cultural dimensions such as individualism-collectivism. Specifically, members of individualistic cultures tend to use more dominating conflict strategies, more substantive, outcome-oriented strategies, and fewer strategies for avoiding conflict than do members of collectivistic cultures, due to higher face concerns among members of collectivist groups.

Hofstede (1980) argued that national culture was the source of a considerable amount of common mental programming of citizens and thus used his dimensions to explore differences between the citizens of various countries (e.g., by designating people from Japan as "collectivist" and people from the U.S. "individualistic"). He argued that national cultural value systems are quite stable over time and can be carried forward from generation to generation. Several of the cross-cultural comparison studies in deception select countries that would be opposite on Hofstede's dimensions (such as Canada-China and Columbia-Australia). However, in pluralistic societies such as the United States, or in societies where there are distinct cultural groups that differ from one another, measuring culture on an individual level rather than according to national identity allows for a more granular analysis.

Recognizing that Hofstede's cultural dimensions are not based on mutually-exclusive traits which exist on opposite ends of a spectrum, others have adapted Hofstede's early measures and revised them so that participants can indicate their identification with both collectivistic and individualistic traits (Singelis et al. 1995). Likewise, items based on horizontalism/verticalism are included in combination with the other dimensions so that different aspects of each construct can be applied, such as vertical collectivism (which values hierarchies within a group) compared to horizontal collectivism (which expects equality among the collective).

Although Hofstede's work is viewed as seminal in the field (Holden 2004), some intercultural communication scholars have criticized Hofstede's nation-level unit of analysis as unsuitable for examining cultural differences (McSweeney 2002). Objections include the fact that culture is likely a far more intricate construct than can be described with five dimensions, the dimensions themselves were originally conceptualized to create what are now understood to be false dichotomies between vague, culturally-loaded concepts such as masculinity and femininity (Jones 2007). Noting that scores on these dimensions could not accurately be considered static, Signorini et al. (2009) explain them to be oversimplifications (see also Yeh 1983 and Wu 2006 for methodological critiques). It should be noted that Hofstede (2002) consistently engaged with his critics, refining his measures and introducing new dimensions to address issues (Hofstede et al. 2010b). Due to the critiques, however, Orr and Hauser (2008) emphasize the importance of collecting supplemental data alongside measuring Hofstede's cultural dimensions, such as including qualitative follow-up questions beyond the self-report cultural dimensions scale items, as well as measuring observed physical behaviors of people from different cultures during interpersonal interactions, both of which were integrated into our SCAN project.

Oetzel and Ting-Toomey (2003) argued for the inclusion of self-construal in measuring culture, by which they mean that researchers should measure the way that individuals conceive of themselves within a larger cultural framework. This is because individuals can vary from the predominant cultural framework of a nation or society, such as those from more interdependent sub-groups within an individualistic culture. "Essentially, cultural values have a direct effect on conflict behaviors and an indirect effect on conflict behaviors that is mediated through individual-level factors" (Oetzel and Ting-Toomey 2003, p. 603). Our SCAN project takes a two-layered approach to studying the influence of culture on deception. We examine differences in deception strategies and behaviors among the nationalities of our participants, but we also ask them to self-report their individual perceptions of their own cultural self-construal. All measures used in this project are outlined in Chap. 11 in this volume by Burgoon et al. (2021); (see also Burgoon et al. 2009), including scales that elaborated upon and expanded Hofstede's initial cultural dimensions from Singelis et al. (1995).

Current Research in Understanding Deception Detection as a Cultural Construct

We begin with aspects of deception that transcend culture. Deception is present across all cultures and impacts every message processed. Although all people lie in some form and in some contexts (Levine et al. 1999; Serota et al. 2010), truth-telling occurs far more frequently in everyday interaction in most contexts (McCornack and Parks 1986). According to the "truth bias" perspective, most people are bad at deception detection simply because most people passively assume that others are telling the truth unless they have some reason for suspicion (Zuckerman et al. 1981). Building from this premise, Levine's (2014) Truth-Default theory suggests that people tend to believe others, and that the truth bias is pro-social and adaptive: "the truth-default enables efficient communication and cooperation, and the presumption of honesty typically leads to correct belief states because most communication is honest most of the time" (Levine 2014, pp. 378–379). Street (2015) echoes these ideas with his adaptive lie detector theory (ALIED), which argues that the truth default or any truth bias is not a fault or a weakness on the part of lie detectors, but instead expectations of honesty are the result of informed and adaptive judgments in situations without significant useful context to judge veracity (see also DePaulo et al. 1996). Given that truth must be generally present for language to be functionally communicative (Grice 1975), assumptions of truth are simply a better guess in most situations than assumptions of dishonesty.

Research further suggests that the rate of deception is not evenly distributed across a given population (Serota et al. 2010; Serota and Levine 2015). Most people do not lie regularly or very often (Serota and Levine 2015) but there are some who do. When excluding "white" lies, there exist only a few prolific liars in any population who engage in deception frequently, but interestingly, they do so much that the number of their lies outpaces the number of truths told by most of the population (Levine 2014). Although the patterns for deception and deception detection described above apply to people across various cultures, research findings on people's motives for deception, deception cues used to detect deception, and understanding the meaning of deception "tells" all reflect cultural differences, as described below.

Motivations for lying Levine et al. (2016) conducted research to track the most common reasons for lying amongst people from different cultures, with the aim of identifying pan-cultural deception motives. Prior research found that lies may be told to benefit the self or others (see DePaulo et al. 1996). The study by Levine et al. (2016) found that participants from Egypt, Guatemala, Pakistan, Saudi Arabia, and the United States all had similar reasons for lying, and that the core human motivation for deception appears to be a desire for personal gain or benefit. Lying for personal gain usually involves obtaining social capital through positive self-impressions or psychological gain (Levine et al. 2016). Other common reasons for why people lie include exercising power over others, maintaining personal privacy, or simple enjoyment (Choi et al. 2011), although it must be recognized that lies can be pro-social, allowing communicative partners to save face (DePaulo et al. 1996; Ekman 1997).

In a study investigating the impact of cultural identity on people's motivations for engaging in deceptive communication, Kim et al. (2008) found that people from more interdependent or collectivistically-oriented groups showed higher overall motivation for lying for both self- and other-benefit. Other studies, however, suggest that people in collectivistic cultures are more likely to lie for others' benefit than are people in individualistic cultures (Triandis et al. 2001). Park et al. (2018) recently found that Koreans were more accepting of lying for protecting a friend than were Americans. The collectivistic value of maintaining harmony among group members was suggested as a possible explanation for these findings. This explanation is supported by other research, including, for example, findings that Americans (individualists) are more likely to lie about issues that are personal, whereas Samoans (collectivists) are more comfortable lying to protect their family or group status (Aune and Waters 1994), and employees in the U.S. are more likely to deceive for personal gain compared to Israeli employees (Sim 2002).

Lying within and between ingroups/outgroups Although the reasons for lying are similar across many cultures, intercultural variation remains important for deception detection in interactions between members of different cultural groups. Nonverbal cues vary between cultures and allow for different heuristics to be used in deception detection. For example, while one of the most commonly-referenced signals for deception detection is eye gaze, which is often (although erroneously) used to determine how honest an individual is (Buller and Burgoon 1994; Global Deception Research Team 2006), Vrij et al. (1992) discovered that while the Dutch market-dominant minority in Suriname consider a lack of eye contact to be very suspicious, the Afro-Dutch in Suriname consider direct eye contact to be a breach of politeness norms. Consequently, individuals from different (sub)cultural groups must either adapt their nonverbal communication strategy or expect miscommunication resulting from intercultural differences. Interestingly, however, despite the near-universal use of eye gaze as an indicator to detect deception, eye gaze fails to provide accuracy in detecting deception across cultural contexts. Moreover, Bond et al. (1990) found that both Jordanians and Americans used different behaviors associated with eye gaze to determine whether an individual is lying or not, and that Jordanians displayed more eye contact than Americans during interactions regardless of whether they had been lying or not. Jack et al. (2012) found that eye gaze was the most significant indicator used for determining honesty for Chinese participants, and yet participants often used contradictory heuristics concerning eye gaze in their honesty evaluations. These cases illustrate that even though eye gaze is a common signal used for deception detection, cross-cultural differences in eye gaze behavior are "much greater than any differences associated with veracity" (Castillo 2015, p. 249).

Research also finds that people tend to treat outgroup members' statements more skeptically than statements from ingroup members (Dunbar et al. 2016; Levine and McCornack 1992; Slessor et al. 2014; Whitty and Carville 2008). This likely stems from the more general phenomenon of *intergroup bias*, which suggests that people prefer those in their ingroup and find them to be more trustworthy than people from outgroups (Hewstone et al. 2002). That said, however, Bond and Atoum (2000) tested intergroup bias in a study of the lie detection abilities of Americans, Jordanians, and Indian nationals. Contrary to expectations, they found that speakers from another culture were not always seen as inherently more suspicious. This suggests that suspicion is more complex than simply a function of cultural differences between interaction partners. Most likely it is instead relative to the specific communicative behaviors displayed by a person from one culture in the context of intercultural interaction. Finally, most research that has examined deception detection across cultures has concluded that no one culture is more adept at detecting deception than others (see Choi et al. 2011; Griffin and Bender 2019; Lapinski and Levine 2000; Levine et al. 2016).

Improperly Using "Tells"   One problem inherent to deception research generally is an overemphasis on nonverbal cues that are unreliable for successful deception detection. Because of this, real-time deception in any interaction is usually accurately detected only slightly above chance at 54% (Aamodt and Custer 2006; Bond and DePaulo 2006; Sporer and Schwandt 2006, 2007). A meta-analysis of deception cues by DePaulo et al. (2003) found over 100 nonverbal cues to deception detection in 120 samples across a wide array of countries. Nonverbal cues can be vocal (e.g., speech hesitations, errors, rate, etc.) or visual (e.g., eye gaze, smile, hand movements, etc.). Moreover, while the meta-analysis reported eye gaze is the most commonly perceived nonverbal cue for deception detection, it found that eye gaze is *not* actually related to deception. Nonverbal cues can be unreliable as deception cues for a variety of reasons: individuals can adapt or modify their behavior during interactions, the motivation or type of lie can affect liar behavior, liars and truth-tellers experience similar stressors and therefore look similar while being questioned, and behaviors have different meanings for different people, both due to individual variation and to cultural differences between interactants. Indeed, intercultural communication brings to the forefront the problem of relying on nonverbal cues in deception detection, as certain patterns of behavior that are associated with dishonesty in one cultural context may not be perceived as suspicious behavior in a different culture (e.g., Vrij et al. 1992).

Properly Understanding "Tells" While nonverbal cues may not work as simple universal "tells" for deception detection, such cues are far from useless when considered in their proper context. Ekman (2001) explains that "there is no sign of deceit itself—no gesture, facial expression, or muscle twitch that in and of itself means that a person is lying. There are only clues that the person is poorly prepared and clues of emotions that don't fit the person's line" or standard interaction style (p. 80). Such clues are indeed significant for detecting deception, but they must be understood in their own unique situated context. Correctly interpreting clues, for example, by placing them in their proper cultural context, shedding unhelpful biases, and eventually building a repertoire of accurate expectations with conversational partners allows for an individual to better detect deception (Frank and Feeley 2003).

To explore this, Vrij (2015) advocates a cognitive approach to deception detection, which asserts that lying requires too much cognitive effort for the deceiver to engage in the conversation with complete fluidity. (For a review of the cognitive approach and prominent voices engaged in this research, see Sporer 2016.) However, recognizing fluidity in others is a particularly difficult skill for outgroup members. Research shows greater success in gauging how fluent a person is if the person is from the same culture (Chen et al. 2002; Hřebíčková and Graf 2013). Thus, cultural competence is an important factor for detecting deception in this approach, as the violation or adherence to cultural norms is opaque to those who are unfamiliar with the cultural context of their conversational partners. Moreover, lying involves generating new imagined possibilities that are close enough to reality to be believed, but do not quite match reality in accordance to the deceiver's goals (Spence 2004), a task which is likely much more difficult cross-culturally. For example, in intercultural communication generally, the repertoire of appropriate strategies for verbal and nonverbal behavior may differ between cultural groups, which means that intercultural deceivers must not simply control their own behavior to appear trustworthy; they must control it in a way that is understood as trustworthy to their interaction partner who may come with a different set of behavioral expectations for truthful communication. Likewise, deception detectors must interpret their partner's behaviors with relevant knowledge of deceptive strategies used in that person's culture in mind. This makes properly understanding deception "tells" in intercultural communication contexts more difficult.

Challenges in Interpreting, Applying, and Integrating Research on Culture and Deception

Deception and deception detection are multidisciplinary areas of focus, spanning a variety of fields, including, for example: communication studies, psychology, sociology, anthropology, philosophy, ethics, law, criminology and forensic science, psychiatry and behavioral neuroscience, counseling, literature, linguistics, business, management, journalism, advertising, public relations, marketing, and political science (Docan-Morgan 2019). As with any interdisciplinary task, coordination among scholars from these groups inevitably involves mismatched lexicons, field-specific jargon, and disparate histories and loaded meanings behind concepts. This does not mean that these different communities cannot productively collaborate, but care must be taken so that conceptual clarity is not lost in translation.

In studying the role of culture in deception, these challenges are exacerbated by the fact that culture is often the secondary topic of interest for researchers, generally seen as moderating the way in which deception occurs. And, as discussed earlier, culture is seldom a simple or precise variable that can be easily isolated or manipulated between groups in a study. As another example, Kim et al. (2008) tested the effects of culture on deceptive motives for participants from Hong Kong versus the United States. The authors argue that the difference they observed between the two groups is explained by the way that collectivists "are willing to stray from the truth if not telling the truth serves to promote harmonious relationships" (p. 42). However, when Levine et al. (2016) examined pan-cultural motives for deception, they found that politeness norms accounted for less than 10% of the variance between the types of lies in collectivist versus individualist societies. Levine et al.'s research did, however, indicate that even if the motives for deception span across cultures, "the situations in which those motives become salient and obstructed by the truth are culturally variable" (p. 4). This brings to the forefront the question of whether there is agreement between participants from different cultures as to what constitutes deception. Some noted constructs that can change the social acceptability and cognizance of what is and is not considered to be lying include the relationship between the deceiver-deceived, the intention of the deceiver, and the cultural context of deception (Seiter et al. 2002).

Misapplying findings cross-culturally Further misunderstanding can occur because often in deception detection research, there is not a clear distinction made between studying the way that individuals behave with members of their own cultural group and how they act in intercultural settings. This is pertinent because if the way that individuals engage in deception is motivated by values specific to their culture (e.g., protecting others' face), they may not engage in the same deceptive strategies when interacting with members of a different cultural group. In theory, more honest communication should happen between cultural ingroup members than between people from culturally distinct groups (Fitch 2010). And lying *is* found to be more common between people from different cultural groups (Knight 1998). This finding has been replicated multiple times in deception literature for over 30 years. As examples, Coleman and Kay (1981) illustrated how English speakers are more likely to be honest with those they consider to be from their ingroup than people from an outgroup, which laid the groundwork for later research illustrating a division between deception strategies based on cultural affiliation (Sweetser 1987). Both Ecuadorians and European-Americans demonstrated the same bias (Mealy et al. 2007), as did university students in the United States (Dunbar et al. 2016). Notably, this effect is more prominent for individuals with collectivistic tendencies

(Fu et al. 2008). Given evidence that cultural tendencies towards collectivism/individualism function differently to affect deception towards ingroup versus towards outgroup members, research examining cultural perspectives on deception must take care to keep intracultural and intercultural deception norms distinct.

Challenge of deception detection contrasted with truth-teller identification When distinguishing between truth and deception, the task of determining credibility (i.e., truth) is just as important as detecting deception. In real-world settings, detecting deception involves more than identifying dishonest statements as false. People are faced with a variety of interactions in day-to-day life in which they may encounter deception, and importantly, the task of successfully detecting deception involves both correctly identifying liars *and* not misidentifying truth-tellers. This is complicated by the fact that culturally-specific cues for distrusting deceivers and trusting truth-tellers are not found at binary extremes – for example, even if an individual believes that gaze aversion signals deception, unbroken eye contact does not necessarily inspire confidence. Truth-tellers do not enact the inverse of deceiver behavior, nor vice versa. Moreover, the signals that allow an individual to determine that someone is trustworthy vary widely across study samples, and there is little consensus about these signals across cultural groups (Hofstede et al. 2010a), but research in this field has been far from comprehensive and continues to grow to fill in these gaps.

For example, in the U.S., police officers are trained to assess credibility based on consistency of statements, contradictions, and level of detail in a subject's verbal responses (Campbell et al. 2015). Elsewhere, more emphasis is given to nonverbal behaviors when judging credibility. For example, whereas smiling in western industrialized societies engenders trust, in societies where corruption is high, smiling works against trust (Krys et al. 2016). Ozono et al. (2010) found that while Japanese participants were more likely to rate strong eye gaze as trustworthy, American participants pay more attention to how much a person smiles. Another study found that Japanese businesspeople emphasized a positive correlation between trustworthiness and level of embeddedness in the group (Nishishiba and Ritchie 2000). Although research generally finds that collectivism motivates increased trust for ingroup members, this is not always the case. Birkás et al. (2014) found that Hungarian participants were more likely to trust other Hungarians than foreigners, while the opposite pattern was common in participants from East and South Asia. Other research finds that trust of members of outgroups increases with greater exposure to those groups (Carney et al. 2007; Heery and Valani 2010; ten Brinke et al. 2014). Put simply, this work suggests that the more an individual has been exposed to people from other cultures, the less likely they are to fall back on oversimplified trust heuristics which are more reliable for judging ingroup members than judging members of outgroups.

These findings indicate that the ways in which trust is built or diminished vary widely across cultural groups. And even if there are some relatively universal heuristics used across cultures, such as those related to eye gaze (Global Deception Research Team 2006), norms surrounding politeness or power distance that are distinct to different cultural groups introduce so much variance that cross-cultural comparisons are easily confounded. Due to these problems, when attempting to engage in intercultural deception detection, an individual must not only discern which statements are lies, they must also (a) be familiar with and (b) apply foreign heuristics to recognize how an individual demonstrates trustworthiness relative to that person's cultural norms.

Challenge of culturally-relative interpretations of truth   While truth is often discussed as a singular and objective concept that different people can agree upon (i.e., there is one real truth), multiple perspectives on truth can and do coexist (see Marwick and Lewis 2017; Zelizer 2009). Across different groups, there exist different beliefs about what truth is, as well as differing qualifications for what constitutes a lie. Given that different groups can disagree about what is and is not true, detecting deception across cultures requires calibration to different cultural epistemologies. Reorienting to disparate cultural ways of knowing is challenging, especially for people from "tight" cultures who expect "peoples' values, norms, and behavior [to be] similar to each other" (Uz 2014, p. 319) versus people from "loose" cultures who have a higher tolerance for deviant behavior (Gelfand et al. 2011).

Looking cross-culturally, the concept of truth is itself fundamentally different in different parts of the world (Jameson 1992; Sweetser 1987), as is the concept of what constitutes a lie (Dor 2017). Lee et al. (2001) found that almost all (87%) of the Canadian participants in their study considered a pro-social deceptive statement to be a lie, whereas only half (52%) of their Chinese participants considered the same statement to be dishonest. Preferences for low- and high-context communication are also significant for determining truth-telling versus deception. Park and Ahn (2007) found that high-context communication was preferred by Korean participants, and only 35% of them found an ambiguous statement to be deceptive, whereas 70% of American participants in their study considered the same statement to be a lie. These examples illustrate the subjectivity of truth as a construct and that people from different cultural groups can subscribe to different notions of truth in the same situation.

From the perspective of someone who expects truth to be singular, the notion of multiple orientations towards truth can be perplexing. However, it is important to note that for individuals from the opposite framework, the notion of a single truth is similarly foreign. Describing doing business across the U.S./Mexican border, Condon (1985) recalls being told:

> You Americans, when you think of a banana, you think of only one kind of fruit. But when you come to Mexico and visit a market, you see that are so many kinds. Some are big and solid and used for cooking, like potatoes. You never heard of such a thing. Others are tiny as your thumb and sweeter than candy. You never imagined such a thing. And I'll tell you, my friend, here in Mexico we have as many kinds of truth as there are kinds of bananas. You don't know what you've been missing. (p. 43)

This quote illustrates an important application of cultural tightness/looseness, which involves recognizing that different epistemologies about truth can coexist (Gelfand 2012). Modern Confucian epistemology likewise places less importance on

distinguishing individual claims as being "true" than it does on how proper, beneficial, or appropriate a claim is (Hall 2001; Hansen 1992). For individuals from
groups with these ideologies (and many other ideologies as well), truth is inherently
associated with a sense of group harmony and responsible stewardship over others.
As such, this cultural worldview posits that a self-serving factual statement can be
less true than a lie that benefits the collective group. Recognizing the importance of
this flexibility, Agar (1994) describes the skills necessary for intercultural communication, recounting that "what we expect and how we define 'the truth' or 'a lie' is
a cultural matter" (p. 228).

Contributions of the SCAN Project

Our work in the SCAN project aims to follow identified best practices to avoid some
of the challenges of cross- and inter-cultural deception research described above.
Following the recommendations of Berry (1980) for evaluating how intercultural
research is done, our research employed an emic/etic approach to studying culture.
The phrases emic and etic are borrowed from linguistics, where phon*emics* describe
the sounds used in a specific language, while phon*etics* constitute all sounds made
across languages. Applying this understanding to deception research that examines
culture, it is essential to break out of one's own individual emic perspective to gain
insight into other groups. As such, cross-cultural researchers are encouraged to recognize how other groups' emic perspectives constitute a valuable piece of etic
knowledge. For example, although our work makes comparisons across cultural
groups, we allow individual participants to define for themselves what deception is
and to use their own notions of trustworthiness in our analyses. Our perspective thus
also follows Gudykunst's (2001) orientation of conducting theoretically-based etic
research that incorporates emic issues where appropriate. We also measured and
controlled for individual-level variables that could explain differences we might
observe across the cultural groups we studied, such as level of English proficiency
and prior knowledge of the game (see van de Vijver and Leung 2000 for a discussion
of important control variables).

This project moreover uses a variety of methods to study deception across cultural groups. Qualitative open-ended questions were asked of participants about the
cues they rely on to detect deception, validated self-report measures were used
throughout the game to gauge participants' perceptions of the other players, and
both audio and video data is being analyzed to identify behavioral and vocalic trends
during game play. This strategy allows for triangulation between different types of
data to better understand the research findings. For example, by comparing liars and
truth-tellers across the globe using self-report and computational analyses of participants' verbal and nonverbal behavior, alongside having ground-truth knowledge
of exactly when a participant is lying or truth-telling, we will be able to evaluate the
importance of (or lack thereof) hypothesized deception cues such as eye gaze, pitch,
turns at talk, or fluidity across different populations, and thus provide support for

competing theories on deceptive behaviors within and across cultures, such as the specific discrimination versus the universal cue perspectives on deception detection.

Conclusion

This chapter demonstrates the complexities of deception and its detection both within and across cultures. Specifically, it sought to illuminate complexities in operationalizing culture in the context of researching deception, in understanding the multitude of roles that culture can play in deception and its detection, for example, in influencing people's motivations for lying in general, for how deception is enacted and detected between members of ingroups and outgroups, and people's use of and ability to understand cues that enable them to lie or to detect lying successfully. It further elucidated challenges for researchers when using a cultural lens to study deception, including the fact that people from different cultures often do not agree on what constitutes deception in the first place, taking care not to overgeneralize findings from intra- to inter-cultural communication contexts (or vice-versa), and placing appropriate focus on identifying both liars *and* truth-tellers within deceptive interaction contexts. Ultimately, it is our hope that the SCAN project, which includes verbal and nonverbal data from six countries spanning five distinct global regions (Asia, North America, Middle East, Africa, and Pacific Islands) will provide the field with one of the most comprehensive views of deception and deception detection available in the literature to date.

Acknowledgement   We are grateful to the Army Research Office for funding much of the work reported in this book under Grant W911NF-16-1-0342.

Funding Disclosure   This research was sponsored by the Army Research Office and was accomplished under Grant Number W911NF-16-1-0342. The views and conclusions contained in this document are those of the authors and should not be interpreted as representing the official policies, either expressed or implied, of the Army Research Office or the U.S. Government. The U.S. Government is authorized to reproduce and distribute reprints for Government purposes notwithstanding any copyright notation herein.

References

Aamodt, M. G., & Custer, H. (2006). Who can best catch a liar? *Forensic Examiner, 15*(1), 6–11.

Agar, M. (1994). *Language shock: Understanding the culture of conversation.* New York: William Morrow and Company.

Al-Simadi, F. A. (2000). Detection of deceptive behavior: A cross-cultural test. *Social Behavior and Personality: An International Journal, 28*(5), 455–461.

Aune, R. K., & Waters, L. L. (1994). Cultural differences in deception: Motivations to deceive in Samoan and North Americans. *International Journal of Intercultural Relations, 19*, 159–172.

Berry, J. (1980). Introduction to methodology. In H. C. Triandis & J. Berry (Eds.), *Handbook of cross-cultural psychology* (Vol. 2, pp. 1–28). Boston: Allyn & Bacon.

Birkás, B., Dzhelyova, M., Lábadi, B., Bereczkei, T., & Perrett, D. I. (2014). Cross-cultural perception of trustworthiness: The effect of ethnicity features on evaluation of faces' observed trustworthiness across four samples. *Personality and Individual Differences, 69,* 56–61. https://doi.org/10.1016/j.paid.2014.05.012

Bond, C. F., & Atoum, A. (2000). International deception. *Personality and Social Psychology Bulletin, 26,* 385–395.

Bond, C. F., & DePaulo, B. M. (2006). Accuracy of deception judgments. *Personality and Social Psychology Review, 10*(3), 214–234.

Bond, C. F., Omar, A., Mahmoud, A., & Bonser, R. N. (1990). Lie detection across cultures. *Journal of Nonverbal Behavior, 14,* 189–205.

Bradac, J. J., Friedman, E., & Giles, H. (1986). A social approach to propositional communication: Speakers lie to hearers. In G. McGregor (Ed.), *Language for hearers* (pp. 127–151). Oxford, UK: Pergamon.

Buller, D. B., & Burgoon, J. K. (1994). Deception: Strategic and nonstrategic communication. In J. A. Daly & J. M. Wiemann (Eds.), *Strategic interpersonal communication* (pp. 191–223). Hillsdale: Erlbaum.

Burgoon, J. K., Levine, T., Nunamaker, J. F., Metaxas, D., & Park, H. S. (2009, August 6). *Rapid noncontact credibility assessment: Credibility assessment research initiative.* Final Report to the Counter-intelligence Field Activity (Contract No. H9C 104-07-C-0011).

Burgoon, J. K., Nunamaker, J. F., & Metaxas, D., & Ge Tina. (2021). Does culture influence deceptive communication? In V. S. Subrahmanian, N. E. Dunbar & J. K. Burgoon (Eds.), *Detecting trust and deception in group interaction.* Springer.

Campbell, B. A., Menaker, T. A., & King, W. R. (2015). The determination of victim credibility by adult and juvenile sexual assault investigators. *Journal of Criminal Justice, 43*(1), 29–39.

Carney, D. R., Colvin, C. R., & Hall, J. A. (2007). A thin slice perspective on the accuracy of first impressions. *Journal of Research in Personality, 41,* 1054–1072.

Castillo, P. A. (2011). *Cultural and cross-cultural factors in judgments of credibility* (Doctoral thesis). Sydney: Charles Stuart University.

Castillo, P. A. (2015). The detection of in cross-cultural contexts. In A. Awasthi & M. K. Mandal (Eds.), *Understanding facial expressions in communication* (pp. 243–263). New Delhi: Springer India.

Chen, Y.-R., Brockner, J., & Chen, X.-P. (2002). Individual-collective primacy and ingroup favoritism: Enhancement and protection effects. *Journal of Experimental Social Psychology, 38,* 482–491.

Choi, H. J., Park, H. S., & Oh, J. Y. (2011). Cultural differences in how individuals explain their lying and truth-telling tendencies. *International Journal of Intercultural Relations, 35,* 749–766.

Coleman, L., & Kay, P. (1981). Prototype semantics: The English word lie. *Language, 57*(1), 26–44.

Condon, J. (1985). *Good neighbors: Communicating with the Mexicans.* Yarmouth: Intercultural Press.

DePaulo, B. M., & Kashy, D. A. (1998). Everyday lies in close and casual relationships. *Journal of Personality and Social Psychology, 74*(1), 63–79.

DePaulo, B. M., Kashy, D. A., Kirkendol, S. E., Wyer, M. M., & Epstein, J. A. (1996). Lying in everyday life. *Journal of Personality and Social Psychology, 70,* 979–995.

DePaulo, B. M., Lindsay, J. J., Malone, B. E., Muhlenbruck, L., Charlton, K., & Cooper, H. (2003). Cues to deception. *Psychological Bulletin, 129,* 74–118.

Docan-Morgan, T. (2019). Preface. In T. Docan-Morgan (Ed.), *The Palgrave handbook of deceptive communication* (pp. v–vi). Cham: Palgrave Macmillan.

Dor, D. (2017). The role of the lie in the evolution of human language. *Language Sciences, 63,* 44–59.

Dunbar, N. E., Gangi, K., Coveleski, S., Adams, A., Bernhold, Q., & Giles, H. (2016). When is it acceptable to lie? Interpersonal and intergroup perspectives on deception. *Communication Studies, 67,* 129–146.

Ekman, P. (1997). Lying and deception. In N. Stein, P. Ornstein, B. Tversky, & C. Brainerd (Eds.), *Memory for everyday and emotional events* (pp. 333–347). New York: Psychology Press.

Ekman, P. (2001). *Telling lies: Clues to deceit in the marketplace, politics and marriage*. New York: W. W. Norton & Company (Originally published 1985).

Fitch, W. T. (2010). *The evolution of language*. New York: Cambridge University Press.

Frank, M. G., & Feeley, T. H. (2003). To catch a liar: Challenges for research in lie detection training. *Journal of Applied Communication Research, 31*, 58–75.

Fu, G., Evans, A. D., Wang, L., & Lee, K. (2008). Lying in the name of the collective good: A developmental study. *Developmental Science, 11*, 495–503.

Gelfand, M. J. (2012). Culture's constraints: International differences in the strength of social norms. *Current Directions in Psychological Science, 21*, 420–424.

Gelfand, M. J., Raver, J. L., Nishii, L., Leslie, L. M., Lun, J., Lim, B. C., et al. (2011). Differences between tight and loose cultures: A 33-nation study. *Science, 332*(6033), 1100–1104.

George, J. F., Gupta, M., Giordano, G., Mills, A. M., Tennant, V. M., & Lewis, C. C. (2018). The effects of communication media and culture on deception detection accuracy. *MIS Quarterly, 42*(2), 551–575.

Giles, H. (Ed.). (2016). *Communication accommodation theory: Negotiating personal relationships and social identities across contexts*. Cambridge, UK: Cambridge University Press.

Global Deception Research Team. (2006). A world of lies. *Journal of Cross-Cultural Psychology, 37*(1), 60–74.

Grice, H. P. (1975). Logic and conversation. In P. Cole & J. Morgan (Eds.), *Syntax and semantics* (Vol. 3, pp. 41–58). New York: Academic.

Griffin, D. J., & Bender, C. (2019). Culture and deception: The influence of language and societies on lying. In T. Docan-Morgan (Ed.), *The palgrave handbook of deceptive communication* (pp. 67–90). Cham: Palgrave Macmillan.

Gudykunst, W. B. (2001). Issues in cross-cultural communication research. In W. B. Gudykunst & B. Mody (Eds.), *Handbook of international and intercultural communication* (pp. 165–177). Thousand Oaks: Sage.

Gudykunst, W. B. (Ed.). (2005). *Theorizing about intercultural communication*. Thousand Oaks: Sage.

Hall, E. T. (1976). *Beyond culture*. New York: Doubleday.

Hall, D. (2001). Just how provincial is Western philosophy? 'Truth' in comparative context. *Social Epistemology, 15*, 285–298.

Hammer, M. R., Gudykunst, W. B., & Wiseman, R. L. (1978). Dimensions of intercultural effectiveness: An exploratory study. *International Journal of Intercultural Relations, 2*(4), 382–393.

Hansen, C. (1992). *A Daoist theory of Chinese thought*. Oxford, UK: Oxford University Press.

Heery, E. A., & Valani, H. (2010). Implicit learning of social predictions. *Journal of Experimental and Social Psychology, 46*, 577–581.

Hewstone, M., Rubin, M., & Willis, H. (2002). Intergroup bias. *Annual Review of Psychology, 53*, 575–604.

Hofstede, G. (1980). Culture and organizations. *International Studies of Management & Organization, 10*(4), 15–41.

Hofstede, G. (2002). Dimensions do not exist – A reply to Brendan McSweeney. *Human Relations, 55*(11), 1355–1361.

Hofstede, G. J., Fritz, M., Canavari, M., & Oosterkamp, E. (2010a). Towards a cross-cultural typology of trust in B2B food trade. *British Food Journal, 112*(7), 671–692.

Hofstede, G., Garibaldi de Hilal, V., Malvezzi, S., Tanure, B., & Vinken, H. (2010b). Comparing regional cultures within a country: Lessons from Brazil. *Journal of Cross-Cultural Psychology, 41*(3), 336–352.

Holden, N. (2004). Why marketers need a new concept of culture for the global knowledge economy. *International Marketing Review, 21*(6), 563–572.

Holliday, A. (2016). Revisiting intercultural competence: Small culture formation on the go through threads of experience. *International Journal of Bias, Identity and Diversities in Education, 1*, 1–14.

Hřebíčková, M., & Graf, S. (2013). Accuracy of national stereotypes in central Europe: Outgroups are not better than ingroup in considering personality traits of real people. *European Journal of Personality, 28*, 60–72.

Jack, R. E., Garrod, O. G. B., Yu, H., Caldara, R., & Schyns, P. G. (2012). Facial expressions of emotion are not culturally universal. *Proceedings of the National Academy of Sciences of the United States of America, 109*(19), 7241–7244.

Jameson, F. (1992). *Postmodernism, or the cultural logic of late capitalism.* Durham: Duke-University Press.

Jones, M. L. (2007). Hofstede – Culturally questionable? *Oxford business & economics conference*, Oxford, UK, 24–26 June 2007.

Kim, M. S., Kam, K. Y., Sharkey, W. F., & Singelis, T. M. (2008). "Deception: Moral transgression or social necessity?": Cultural-relativity of deception motivations and perceptions of deceptive communication. *Journal of International and Intercultural Communication, 1*, 23–50.

Knight, C. (1998). Ritual/speech coevolution: A solution to the problem of deception. In J. R. Hurfurd, M. Studdert-Kennedy, & C. Knight (Eds.), *Approaches to the evolution of language: Social and cognitive bases* (pp. 68–91). Cambridge, UK: Cambridge University Press.

Krys, K., Vauclair, C. M., Capaldi, C. A., Lun, V. M. C., Bond, M. H., Domínguez-Espinosa, A., et al. (2016). Be careful where you smile: Culture shapes judgments of intelligence and honesty of smiling individuals. *Journal of Nonverbal Behavior, 40*(2), 101–116.

Lapinski, M. K., & Levine, T. R. (2000). Culture and information manipulation theory: The effects of self-construal and locus of benefit on information manipulation. *Communication Studies, 51*, 55–73.

Leal, S., Vrij, A., Vernham, Z., Dalton, G., Jupe, L., Harvey, A., & Nahari, G. (2018). Cross-cultural verbal deception. *Legal and Criminological Psychology, 23*, 192–213.

Lee, K., Cameron, C. A., Xu, F., Fu, G., & Board, J. (1997). Chinese and Canadian children's evaluations of lying and truth-telling. *Child Development, 64*, 924–934.

Lee, K., Xu, F., Fu, G., Cameron, C. A., & Chen, S. (2001). Taiwan and Mainland Chinese and Canadian children's categorization and evaluation of lie- and truth-telling: A modesty effect. *British Journal of Developmental Psychology, 19*(4), 525–542.

Levine, T. R. (2014). Truth-default theory (TDT). *Journal of Language and Social Psychology, 33*, 378–392.

Levine, T. R., & McCornack, S. A. (1992). Linking love and lies: A formal test of the McCornack and Parks model of deception detection. *Journal of Social and Personal Relationships, 9*, 143–154.

Levine, T., Park, H., & McCornack, S. (1999). Accuracy in detecting truths and lies: Documenting the "veracity effect". *Communication Monographs, 66*, 125–144.

Levine, T. R., Ali, M. V., Dean, M., Abdulla, R. A., & Garcia-Ruano, K. (2016). Toward a pan-cultural typology of deception motives. *Journal of Intercultural Communication Research, 45*, 1–12.

Lowry, P. B., Zhang, D., Zhou, L., & Fu, X. (2010). Effects of culture, social presence, and group composition on trust in technology-supported decision-making groups. *Information Systems Journal, 20*(3), 297–315.

Marwick, A., & Lewis, R. (2017). *Media manipulation and disinformation online [whitepaper].* New York: Data & Society Research Institute.

Matsumoto, D., Kudoh, T., & Takeuchi, S. (1996). Changing patterns of individualism and collectivism in the United States and Japan. *Culture & Psychology, 2*(1), 77–107.

McCornack, S. A., & Parks, M. R. (1986). Deception detection and relationship development: The other side of trust. *Annals of the International Communication Association, 9*, 377–389.

McLuhan, M. (1970). Education in the electronic age. *Interchange, 1*(4), 1–12. Retrieved 7/5/2019 from https://link.springer.com/content/pdf/10.1007%2FBF02214876.pdf.

McSweeney, B. (2002). The essentials of scholarship: A reply to Geert Hofstede. *Human Relations, 55*(11), 1363–1372.

Mealy, M., Stephan, W., & Urrutia, I. C. (2007). The acceptability of lies: A comparison of Ecuadorians and Euro-Americans. *International Journal of Intercultural Relations, 31*, 689–702.

Newmark, E., & Asante, M. K. (1976). *Intercultural communication.* Retrieved from ERIC Database (ED125009).

Nishishiba, M., & Ritchie, D. (2000). The concept of trustworthiness: A cross-cultural comparison between Japanese and U.S. businesspeople. *Journal of Applied Communication Research, 28*(4), 347–367.

Oetzel, J. G., & Ting-Toomey, S. (2003). Face concerns in interpersonal conflict: A cross-cultural empirical test of the face negotiation theory. *Communication Research, 30*(6), 599–624.

Orr, L. M., & Hauser, W. J. (2008). A re-inquiry of Hofstede's cultural dimensions: A call for the 21st century. *Marketing Management Journal, 18*(2), 1–19.

Ozono, H., Watabe, M., Yoshikawa, S., Nakashima, S., Rule, N. O., Ambady, N., & Adams, R. B. (2010). What's in a smile? Cultural differences in the effects of smiling on judgments of trustworthiness. *Letters on Evolutionary Behavioral Science, 1*, 66–72.

Park, H. S., & Ahn, J. Y. (2007). Cultural differences in judgment of truthful and deceptive messages. *Western Journal of Communication, 71*(4), 294–315.

Park, H. S., Choi, H. J., Oh, J. Y., & Levine, T. R. (2018). Differences and similarities between Americans and Koreans in lying and truth-telling. *Journal of Language and Social Psychology, 37*, 562–577.

Sarbaugh, L. E. (1979). *Intercultural communication.* New Brunswick, NJ: Transaction Publishers.

Seiter, J. S., Bruschke, J., & Bai, C. (2002). The acceptability of deception as a function of perceiver's culture, deceiver's intention, and deceiver-deceived relationship. *Western Journal of Communication, 66*, 158–180.

Serota, K. B., & Levine, T. R. (2015). A few prolifc liars: Variation in the prevalence of lying. *Journal of Language and Social Psychology, 34*, 138–157.

Serota, K. B., Levine, T. R., & Boster, F. J. (2010). The prevalence of lying in America: Three studies of self-reported lies. *Human Communication Research, 36*(1), 2–25.

Signorini, P., Wiesemes, R., & Murphy, R. (2009). Developing alternative frameworks for exploring intercultural learning: A critique of Hofstede's cultural difference model. *Teaching in Higher Education, 14*(3), 253–264.

Sim, R. L. (2002). Support for the use of deception within the work environment: A comparison of Israeli and United States employee attitudes. *Journal of Business Ethnics, 35*, 27–34.

Singelis, T. M., Triandis, H. C., Bhawuk, D. P., & Gelfand, M. J. (1995). Horizontal and vertical dimensions of individualism and collectivism: A theoretical and measurement refinement. *Cross-Cultural Research, 29*(3), 240–275.

Slessor, G., Phillips, L. H., Ruffman, T., Bailey, P. E., & Insch, P. (2014). Exploring own-age biases in deception detection. *Cognition & Emotion, 28*(3), 493–506.

Soares, A. M., Farhangmehr, M., & Shoham, A. (2007). Hofstede's dimensions of culture in international marketing studies. *Journal of Business Research, 60*(3), 277–284.

Spence, S. A. (2004). The deceptive brain. *Journal of the Royal Society of Medicine, 97*, 6–9.

Sporer, S. L. (2016). Deception and cognitive load: Expanding our horizon with a working memory model. *Frontiers in Psychology, 7*, 1–12.

Sporer, S. L., & Schwandt, B. (2006). Paraverbal indicators of deception: A meta-analytic synthesis. *Applied Cognitive Psychology: The Official Journal of the Society for Applied Research in Memory and Cognition, 20*(4), 421–446.

Sporer, S. L., & Schwandt, B. (2007). Moderators of nonverbal indicators of deception: A meta-analytic synthesis. *Psychology, Public Policy, and Law, 13*(1), 1–34.

Street, C. N. (2015). ALIED: Humans as adaptive lie detectors. *Journal of Applied Research in Memory and Cognition, 4*, 335–343.

Sweetser, E. (1987). The definition of lie. In D. Holland & N. Quinn (Eds.), *Cultural models in language and thought* (pp. 43–66). Cambridge, UK: Cambridge University Press.

Taylor, P. J., Larner, S., Conchie, S. M., & Menacere, T. (2017). Culture moderates changes in linguistic self-presentation and detail provision when deceiving others. *Royal Society Open Science, 4*, 1–11.

ten Brinke, L., Stimson, D., & Carney, D. R. (2014). Some evidence for unconscious lie detection. *Psychological Science, 25*(5), 1098–1105.

Ting-Toomey, S. (1988). Intercultural conflict styles: A face negotiation theory. In Y. Kim & W. Gudykunst (Eds.), *Theories in intercultural communication* (pp. 213–235). Newbury Park: Sage.

Ting-Toomey, S., Yee-Jung, K. K., Shapiro, R. B., Garcia, W., Wright, T. J., & Oetzel, J. G. (2000). Ethnic/cultural identity salience and conflict styles in four US ethnic groups. *International Journal of Intercultural Relations, 24*(1), 47–81.

Triandis, H. C., Carnevale, P., Gelfand, M., Robert, C., Wasti, A., Probst, T., et al. (2001). Culture and deception in business negotiations: A multilevel analysis. *International Journal of Cross-Cultural Management, 1*, 73–90.

Uz, I. (2014). The index of cultural tightness and looseness among 68 countries. *Journal of Cross-Cultural Psychology, 46*(3), 319–335.

van de Vijver, F., & Leung, K. (2000). Methodological issues in psychological research on culture. *Journal of Cross-Cultural Psychology, 31*, 33–51.

Vrij, A. (2015). A cognitive approach to lie detection. In P. A. Granhag, A. Vrij, & B. Verschuere (Eds.), *Deception detection: Current challenges and new approaches* (pp. 205–229). Chichester: Wiley.

Vrij, A., & Winkel, F. W. (1991). Cultural patterns in Dutch and Surinam nonverbal behavior: An analysis of simulated police/citizen encounters. *Journal of Nonverbal Behavior, 15*(3), 169–184.

Vrij, A., Dragt, A., & Koppelaar, L. (1992). Interviews with ethnic interviewees: Nonverbal communication errors in impression formation. *Journal of Community & Applied Social Psychology, 2*, 199–208.

Whitty, M. T., & Carville, S. E. (2008). Would I lie to you? Self-serving lies and other-oriented lies told across different media. *Computers in Human Behavior, 24*(3), 1021–1031.

Wu, M. (2006). Hofstede's cultural dimensions 30 years later: A study of Taiwan and the United States. *Intercultural Communication Studies, 15*, 33–42.

Yager, M., Strong, B., Roan, L., Matsumoto, D., & Metcalf, K. A. (2009). *Nonverbal communication in the contemporary operating environment (Project #622785.A790)*. Arlington: U.S. Army Research Institute for the Behavioral and Social Sciences. Retrieved 7/5/2019 from https://apps.dtic.mil/dtic/tr/fulltext/u2/a501219.pdf.

Yeh, R.-S. (1983). On Hofstede's treatment of Chinese and Japanese values. *Asia Pacific Journal of Management, 6*(1), 149–160.

Zelizer, B. (2009). *The changing faces of journalism: Tabloidization, technology and truthiness*. New York: Routledge.

Zuckerman, M., DePaulo, B. M., & Rosenthal, R. (1981). Verbal and nonverbal communication of deception. In L. Berkowitz (Ed.), *Advances in experimental social psychology* (Vol. 14, pp. 1–59). New York: Academic.

Part II
The SCAN Project

Chapter 4
A System for Multi-person, Multi-modal Data Collection in Behavioral Information Systems

Bradley Dorn, Norah E. Dunbar, Judee K. Burgoon, Jay F. Nunamaker, Matt Giles, Brad Walls, Xunyu Chen, Xinran (Rebecca) Wang, Saiying (Tina) Ge, and V. S. Subrahmanian

Considerations for Extending Dyadic Deception Research to Group Deception Research

Analyzing group communication for deception cues involves a few key types of changes. In groups, there are more people to observe, different sender-receiver relationships, and different task structures. In addition to simply requiring more of the same methods, these changes introduce some new challenges. To discuss these new challenges, we will first provide a list of assumptions that are typically made in dyadic research that may or may not apply to group research:

1. The intended target of all communications is singular and known
2. Deceivers are capable of observing and responding to suspicion signals from the other singular communicator
3. For successful deception, a deceiver must only convince one person
4. Only one conversation is occurring at a time
5. Relationships of background information between communicators are homogeneous
6. Deceivers are entirely responsible for implementing their deceptions

In the next section of this chapter, we discuss these assumptions, the possibility of maintaining these assumptions and implications of violating them.

B. Dorn · J. K. Burgoon · J. F. Nunamaker · B. Walls · X. Chen · X. (R.) Wang · S. (T.) Ge
University of Arizona, Tucson, AZ, USA
e-mail: jnunamaker@cmi.arizona.edu

N. E. Dunbar (✉) · M. Giles
University of California, Santa Barbara, Santa Barbara, CA, USA
e-mail: ndunbar@ucsb.edu

V. S. Subrahmanian
Dartmouth College, Hanover, NH, USA

© Springer Nature Switzerland AG 2021   57
V. S. Subrahmanian et al. (eds.), *Detecting Trust and Deception in Group Interaction*, Terrorism, Security, and Computation,
https://doi.org/10.1007/978-3-030-54383-9_4

Assumption 1: The Intended Target of all Communications Is Singular and Known

In dyadic communication, one person is communicating with one other person. This means that there is only one possible target for communication and potential deception. With group communication, it becomes possible for side conversations or conversations that are only relevant for a particular subset of group members. This creates a problem of potentially not knowing the intended target of communication signals. For example, if a group has multiple deceivers, and a message is communicated only between deceivers to facilitate deception, that communication is likely to be no longer deceptive. Ultimately, if one desires to analyze communications at the individual or dyad level, we must have a way of defining the intended target of communications.

Some messages are obviously directed. For example, text messages delivered by computer to a single individual can be recorded in a way that maintains which messages were sent by which individual to another individual. Without a known restriction on which individuals participated in dyad-level communications, it can be difficult to define the extent to which each other individual is an intended target of communication. For example, if one person says something, vocally, intended for one particular person in a group, it is possible that every member of the group hears this message. We may also expect that the person speaking understands that everyone in the group can hear this message. In this case, we argue that the entire group is the actual communication target, even though the message may only be particularly relevant for a subset of group members.

For this reason, we recommend that in observing group communication, directed communications must be distinctly isolated, and communications observable by the entire group or subgroup be treated as though all participating individuals are relevant communication targets.

Assumption 2: Deceivers Are Capable of Observing and Responding to Suspicion Signals from the Other Singular Communicator

According to Interpersonal Deception Theory, deception is an interactive process in which a deceiver responds to a deception target, updating behavior based on communication signals, including suspicion. In the dyadic case, the single deception target has a single level of suspicion about the deceptive communication. In a group context, different individuals in the group may have differing levels of suspicion.

Assumption 3: For Successful Deception, a Deceiver Must Only Convince One Person

With dyadic deception, a deceiver only needs to convince one receiving party of the desired false impression. When a group is involved, each member of the group is relevant to the extent that they impact the desired outcome from deception. When the deception is deception for deception-sake (i.e., it only matters that the other party is deceived, not that the deception is useful for anything), it becomes most similar to dyadic deception. When a secondary goal is the true purpose of deception, for example convincing a group of a more favorable decision outcome based on false information, members of the group become relevant targets of deception to the extent that they can influence that decision or uncover the deception on behalf of those that make this decision.

Assumption 4: Only One Conversation Is Occurring at a Time

While this may not be strictly true in dyadic communication (for example, consider the case where two individuals are talking over each other about different topics), any group communication where backchanneling is feasible can result in multiple completely independent conversations with independent goals occurring at the same time.

Assumption 5: Relationships of Background Information Between Communicators Are Homogeneous

In dyadic research, all communicators are from a similar cultural background, or all communicators are from different cultural backgrounds; there is not a combination of similar and differing in one conversation. With group research, it is possible that one subgroup shares background characteristics that differ from another subgroup.

Assumption 6: Deceivers Are Entirely Responsible for Implementing Their Deceptions

Dyadic theories of deception consider how one deceiver interacts with a deception target. Certain aspects of deception, like added cognitive load, may become more relevant as cognitive demands of handling more complex communication further increase, while others, like guilt, may decrease from the sense of duty to the group for which an individual is participating in the deception.

Research Design Decisions

One of the first decisions to make in designing a deception detection study is how one will obtain or infer ground truth. To simplify this process, we opted to base our deception manipulation on an established hidden identity game where a subset of the participating individuals is incentivized to deceive the others about their identity as members of that team. After reviewing a variety of hidden identity games, we decided that the Resistance by Don Eskridge was the closest to what would be suitable for our experiment.

After establishing the Resistance as the base set of rules to ensure deception incentivization, we modified the game, with repeated rounds of testing, to better capture other behaviors of interest. In this specific study, between-individual dominance, liking, and trust were also of particular interest. To build a baseline of these behaviors, independent of any deceptive-incentive influence, we measured these attitudes after an ice-breaking activity that we included to promote more open interaction between participants. We also modified the Resistance to incorporate a leader election phase so that players would specifically campaign to be in charge, where the group decided which player they wanted to be the leader, as opposed to a systematic progression of which player was the leader in the original game. We also added a hand-raising process, prompting individuals to indicate their votes to the group in addition to secret votes for approval used in the base game.

In addition to these game design changes, we also designed an experiment implementation process to ensure that good data could be collected efficiently. Because participants unfamiliar with this type of game often struggled to understand the game during our practice sessions, we made many revisions to the experiment script for clarity and added a practice round to test for understanding and encourage participants to ask for clarity on parts of the experiment they did not understand. We also found that participants would often move out of the camera frame unless we specifically instructed them to stay in frame and provided them with visual feedback of their position within the frame. We tested for pacing with repeating survey measures to ensure participants did not forget about their position within the game while completing the survey for a pause before returning to it. We also found that card distribution for communicating secret information was inefficient, consuming too much time for participants to continue to be engaged, so we automated as much as we could with software and participant computers with privacy screens.

The full experiment will be discussed in the experiment manual section of this chapter. In the next section, we will talk about our implementation, sample characteristics, and specifics to our implementation.

Our Experimental Method and Implementation

In order to examine deceivers' strategies and truth-tellers' deception detection abilities, we set up this game, a version of the *Resistance* game that was played in a face-to-face context. The same procedures were implemented at eight universities across six countries.

Participants

Participants ($N = 693$; $M_{age} = 22$, $SD_{age} = 3.75$) were primarily college students, although some participants were recruited from the general public. Data collection took place at 8 public universities in the Southwestern US (9 games; $n = 59$), Western US (11 games; $n = 67$), Northeastern US (10 games; $n = 74$), Israel (10 games; $n = 71$), Singapore (12 games; $n = 84$), Fiji (14 games, $n = 106$), Hong Kong (15 games, $n = 115$), and Zambia (15 games, $n = 117$). Participants were recruited via email and advertisements on public message boards. The sample was 59% female and was ethnically diverse (although this varied by location), and the biggest groups were Asian (38%) and White (18%). They reported nationalities representing 41 different countries. Participants were required to be proficient English speakers.

Procedure

Participants signed up for an experiment session using an online scheduling system. The sessions ranged from five to eight participants. After signing up, participants were sent an email with a unique identifier number and a link to a pre-survey, which included consent forms, cultural measures, and demographic questions. Upon arrival at the lab, participants were randomly assigned to one of eight computer stations that included a desk, a tablet with a built-in webcam, and a chair. Participants were given instructions for the game and were informed that they would be filmed by several cameras (e.g., overhead camera, webcam).

After all participants were seated, the experiment facilitator explained the rules of the game, and participants took part in an ice-breaker activity to get to know the other players. After this ice-breaker activity, players rated each other on several scales. Participants took part in the game for an hour, during which they played between three and eight rounds. After the second, fourth, and sixth rounds, and at the end of the game, participants filled out several scales about their attitudes about the other players. Participants were paid for participating, and there were small financial incentives for performing well in the game.

Gameplay

We pilot tested several versions of the game and adapted the rules so as to best address the research questions. Players were randomly and secretly assigned to play deceivers (called "Spies"), or truth-tellers (called "Villagers"). In games of five or six players, two were assigned to be Spies, and in games of seven or eight players, three were assigned to be Spies. The Spies were aware of who the other Spies were, but the villagers did not know anyone else's role. Villagers had to depend on shared information to deduce the other players' identities within the game.

Players completed a series of "missions" by forming teams of varying sizes. At the beginning of each round, players elected a leader, who then chose other players for these missions based on who they thought would help them win the game. All players voted to approve or reject the team leader and then voted on the leader's proposed team. Players voted secretly on their computer, and also voted publicly by raising their hands. They were allowed to vote differently on the computer than in public, and the facilitator would announce if there were a discrepancy in public and private votes, letting participants know that deception had occurred. Those who were chosen by the leader to go on the mission team secretly voted for the mission to succeed or fail. Villagers won rounds by figuring out who the spies were and excluding them from the mission teams. Spies won rounds by causing mission failures. The ultimate winters of the game (Spies or Villagers) were determined by which team won the most rounds. Additionally, players won monetary rewards by being voted as a leader or winning the game.

Measures

Dependent Measures

Game outcome In the Zhou et al. (2013) Mafia study, they operationalized deception detection success as the truth-tellers winning the game (i.e., if the truth-tellers win, they must have accurately detected deception). Similarly, in this study, game outcome was a dichotomous variable measuring whether or not Spies or Villagers won the game.

Trust The extent to which participants trusted each of the other players was measured using a single-item repeated measure, which was asked after the ice-breaker and then every even-numbered round during the game. The item read: *Please rate how much you **trust** each player. Are they trustworthy or suspicious? A rating of 5 would mean they seem honest, reliable and truthful and 1 would mean you thought they were dishonest, unreliable and deceitful* (1 *Not at all* to 5 *Very* much; M = 3.29, SD = 1.36). Because participants responded to this item three to five times about each of the other players, we chose to use a single item in order to avoid fatigue.

Covariates and Moderators

Dominance The extent to which participants found other players to be dominant was measured using a single-item repeated measure (after the ice-breaker and each of the even-numbered rounds). Participants read the following text: *Please rate how dominant each player is. Are they active and forceful or passive and quiet? A rating of 5 would mean you thought they were assertive, active, talkative, and persuasive. A score of 1 would mean you thought they were unassertive, passive, quiet and not influential. Please mark any number from 1 to 5* (1 *Not at all* to 5 *Very* much; $M = 3.28$, $SD = 0.87$).

Previous game experience Participants' previous experience playing similar games was evaluated after the completion of the post-game measures. Participants indicated that they had or had not played a similar game. In this study, 54.2% said they had not played a similar game before.

Collectivism Though it is possible to label certain countries as more or less collectivistic (Hofstede, 1983), our samples involved many participants who were not originally from the country where the data were collected. Therefore, we decided it was appropriate to measure individual differences in the extent to which participants subscribed to collectivism. This was measured using a shortened 5-item version of the Singelis et al. (1995) horizontal collectivism scale (e.g., *I feel good when I cooperate with others*) (1 *Not at all* to 7 *Very much*; $M = 5.77$, $SD = .77$; $\alpha = .79$).

Experiment Manual

In the following section, we will describe how we implemented this group deception experiment with 5–8 players. Though groups of larger numbers of players are likely to be possible, fewer than 5 players are only possible with a single spy, and fewer than 3 players are not possible.

Step1. Procedure Explanation and Ice-Breaker

Ice-Breaker In our experience, participants engage more with the game when they are more comfortable communicating with one another. For this reason, incorporating an ice-breaker activity closer to the beginning may result in a more interactive session. However, this step is not strictly necessary. One example ice-breaker activity, which is the activity we used, is to have each participant say a bit about their background, including an interesting, memorable fact. Then we assigned another participant to ask a follow-up question about that participant's interesting fact.

Explain Rules After the (possible) ice-breaker activity, the rules of the game must be explained to the participants. During this set of instructions, we include specifying the number of spies in the game (though not their identity yet), the number of rounds (or time limit), and compensation for completing various actions in the game (winning, losing, and being elected leader were used in our implementation).

Practice Round Once the rules have been communicated to the participants, we (optionally) recommend conducting a practice round. During this round, an arbitrary set of spies are selected (i.e. not necessarily the players that will be spies during the game), and everyone in the session has full knowledge of which players are spies (which is not the case during the real game). Then participants will complete a full round of the experiment, using public voting means (e.g. raising hands) instead of private voting means. During each phase of the round (leader selection, team selection, mission), the experiment facilitator should discuss what would have happened if the votes had gone differently. In our experience, the practice round brings many opportunities to clarify confusion and for participants to raise questions about parts of the game they don't understand.

Step 2. Assign Game Roles

Determine the number of spies For game balance, the number of spies in a game depends on the number of participants in the session. For 5 or 6 player games, we recommend 2 spies; for 7 or 8 player games, we recommend 3 spies. Table 4.1 below shows how many players we recommend using for a game of each size.

Communicate Role Information to Participants To assign and communicate roles, you will need 3 things: (1) a randomization technique, (2) a method for private communication to participants, (3) a way for spies to know who the other spies are. For efficiency and consistency, we developed custom software that automated these processes. If using a software approach, role information can be communicated to participants as well as information about which other players are also spies to the spies.

In situations where computer-facilitated options are not available, a deck of cards and a "night phase" are also feasible. To complete this approach, draw 6–8 (depending on players) cards, the first 2–3 (depending on number of spies) will be spy cards, the rest, villager cards. Write down which cards represent which roles. Shuffle these cards, then deliver 1 card to each player, face-down. Have players privately look at their cards and announce which card represents which role (e.g. 2 of hearts is a spy). Then have every player close their eyes, face down. Then ask only the spies to open

Table 4.1 Number of Agents recommended for each game size

Number of players	5	6	7	8
Number of spies	2	2	3	3

their eyes and identify which other players are spies (this is the "night phase"). Then have everyone return to closing their eyes, face down, before everyone returns to opening their eyes, face up.

Step 3. Complete Rounds

3.1. Round Introduction   At the beginning of each round, information about how the round will be played is communicated to the participants. The key pieces of information are the current score, the number of players that will be selected for the mission this round, and the number of fail votes required to cause a mission to fail. The round size will vary depending on the number of players in the game.

The tables below show team size (Table 4.2) and the number of fail votes (Table 4.3) required for mission failure for each of the group size and rounds we used.

If using any form of audio-video recording, we recommend playing a synchronization message. In our implementation, we used a high-pitched bell-like "ding" noise to synchronize audio from multiple sources.

3.2. Leader Selection   After announcing important information about the upcoming round, the leader election for the new round begins. For this portion of the round, we conducted both a hand-raise (public vote to signal to the group) and a private vote (recorded by computer voting software). Players can nominate any player, including themselves. Once nominated, players vote to elect the nominated player as the leader of this round. Leaders are elected with majority approval. We opted to provide additional compensation for players elected leader. To prevent single players from dominating the entire game, individual players cannot be leaders

Table 4.2 Team size for missions for each of the group size and rounds

Team size (including leader)								
Number of players	Round 1	Round 2	Round 3	Round 4	Round 5	Round 6	Round 7	Round 8
6	3	3	4	4	4	5	5	5
7	3	3	4	4	5	4	4	4
8	3	3	4	4	5	5	5	5

Table 4.3 Number of fail votes required for mission failure for each of the group size and rounds

Number of fail votes								
Number of players	Round 1	Round 2	Round 3	Round 4	Round 5	Round 6	Round 7	Round 8
6	1	1	1	1	1	2	2	2
7	1	1	1	2	2	2	2	2
8	1	1	1	2	2	2	2	2

in back-to-back rounds. The leader is responsible for proposing collections of players to the group to be voted upon (majority approval). Though many proposals for leaders may occur in succession, we recommend allowing the group to develop methods for selecting which player to vote upon (wait to intervene until discussion has completed). If consensus about which player should be nominated is not reached naturally, create a list of nominated players and conduct individual votes for approval, starting with the first player nominated leader. The first player to receive majority approval will be the leader for that round. Other players on the list will need to be nominated again for consideration to be elected in subsequent rounds.

3.3. Team Selection Once the leader has been elected for the round, the leader will select a number of players to join them on the mission for the round. This set of players, along with the leader is a proposed mission team for the round. It is important to give all members of the group time to speak. Though the leader is ultimately selecting the team for approval voting, all members of the group are able to discuss a leader's selection before voting.

Once a team is selected, the group will then vote, using procedures similar to the leader vote to approve or reject the team. If a majority of votes from all players are received to approve the team, the team will go on to submit mission votes. If the team is rejected, the leader will select another set of players for voting. If 3 teams are rejected, the mission fails, awarding a point to the spies, and the round is ended.

3.4. Mission The mission is the point in the game that decided which team is awarded points. Using entirely private voting procedures (no public mission votes are applicable), the players on the approved team submit "success" or "fail" votes. Though villagers will always want the mission to succeed, spies may want to submit success votes to conceal their identity.

3.5. Round Conclusion Once all votes have been received by the mission team members, the number of "success" votes and "fail" votes are revealed to the group (though not which players submitted each vote). For example, if 3 fail votes are submitted on a team of size 3, the group will know that all 3 players on the mission team submitted a fail vote. After revealing the number of each type of vote, announce the outcome, based on the number of required fail votes for failing the mission and an updated score based on the outcome of this round. After announcing the outcome, give players some time to discuss why they think the mission resulted in the way it did.

3.6. Implemented Script Example Below, in Fig. 4.1, we have provided the general script we used for our national and international data collection. This includes specific notes to facilitators for performing actions as well as dialog options used to promote additional discussion when groups were not sufficiently communicative. This does not include our description of the rules to participants, the practice round, the ice-breaker activity, discussion of compensation, concluding, or debriefing. This is only the general form of what was stated by the facilitator every round.

Round [X]

This is the beginning of round [X].

> Play synchronization media

Leader

In this round, the leader will select [#] players for a mission, and [#] fail vote(s) will cause the mission to fail, resulting in one point for the spies. Who would like to nominate someone else or themselves to be the leader of the first mission?

---- Wait for nomination ----

- *If other nom: For nominator:* **Why would this person make a good leader?**
  For nominated: **Why should you be leader?**
- *If self nom:* **Why would you like to be the leader of this mission?**

Is there discussion of this choice?

---- Wait for discussion ----

Let's take a vote: All in favor of <Player> as the leader, please record your vote. Raise your hands if you voted for <Player>.

> Conduct leader vote procedures using software, cards, or hands.

---- Wait until votes submitted ----

The vote passed/failed, with _____ votes in favor and _____ votes against. *If the vote failed:* **"That vote failed. Player #(X) cannot be the leader for this round. Who do you think should be the leader this round instead?**

> Record leader election outcome

- *If secret recorded vote doesn't match public voting procedures, use this question whenever you need to spur discussion: (Possible) Whip around:* **Those votes did not match up – what do you folks think happened?** *(if they don't give a reason, ask "why?")*
- *[if not a majority, repeat the process]: Remind:* **If there are three failed votes, the spies win the round**

Fig. 4.1 Implemented Script Example

Team
_____

Leader <Player>, would you please select [#] other players to join you on this mission.

---- Wait team-member selection ----

For each team member selected: **Why would you like this person on the mission team?**

OK, you can now discuss this choice before we take a vote on approving this team. If a team fails to receive majority approval, you must try again. A tie counts as a failure. If after 3 tries, you cannot form a team, this counts as a mission failure and gives a point to the Spies.

All in favor of this team, <Players X, Y and Z >? Please record your vote on your computer. Raise your hands if you voted for this team.

Conduct team voting procedures

---- Wait for votes -----

The vote passed/failed, with _____ votes in favor and _____ votes against

- *If secret vote doesn't match public vote: (Possible) Whip around:* **Those votes did not match up – what do you folks think happened?**
- *If not a majority, repeat the process, Remind:* **If there are three failed votes, the spies win the round**

Any discussion?

---- Wait for discussion ----

Possible Questions: **If you had to guess which players are spies, who seems suspect?**

Fig. .4.1 (continued)

Step 4. Complete Surveys

Depending on the goals of the research, it is important to have participants complete surveys at regular intervals. In our implementation, we had participants complete a survey before arriving and after the game, but before they left.

The more difficult type of information to gather is attitudinal information that changes as the game progresses. While it may be necessary to collect this

Mission

OK, the team members now get to decide whether to vote for a mission success or mission failure succeed or fail the mission. TEAM MEMBERS: A box will appear on your screen where you choose to succeed or fail the mission. Please record your vote now. I will reveal the results.

Record vote outcome

---- Wait for votes -----

[#] *FAIL VOTE(S) FAILS THE MISSION:* **We have one (success/fail), and one (success/fail), and one (success/fail)…. So, this mission: SUCCEEDED!/FAILED!, so <Congratulation/Condolences> to the Villagers.**

---- Wait for discussion ----

Fig. .4.1 (continued)

information mid-game, collecting this information can also increase time requirements for experiment implementation and disrupt gameplay, resulting in decreased or unnatural communication. In our implementation, we settled on a re-surveying of attitudinal information after every 2 rounds of gameplay, for up to 5 total evaluations, including 1 evaluation before round 1, after round 2, after round 4, after round 6, and after round 8.

Though differing groups are likely to vary widely in how much time they require to complete rounds, particularly in later rounds, it is not fair to participants to run excessively over estimated time requirements. Further, participants will start to change gameplay, as they notice that more time than they allocated has passed, particularly if they scheduled other obligations around estimated times. One technique to balance this issue is to set a guaranteed time window. In our implementation we decided that the game would end after roughly 1 hour of gameplay, regardless of the number of rounds completed. This results in differing amounts of data available and creates challenges in data collection. It also provides an opportunity for participants to intentionally delay rounds when they are ahead to limit the chance that the other team will gain points. To be able to make sure that a consistent number of rounds (and survey observations) occur, we recommend keeping a round timer on a round to round basis, and make sure that each of 8 rounds are completed within the time estimate provided to participants. Although 1 hour is more than enough for some groups for gameplay, it is not nearly enough for others without effort to force the group through the game more quickly.

Step 5. Conclude and Debrief

Once the experiment is completed, it is important to have participants complete any remaining survey instruments that require that they do not have unintended knowledge before participants reveal that information to each other. Players will be tempted to reveal their roles once gameplay has ended, so before the final round outcome is announced, make sure to tell players when it is appropriate to reveal their roles. In our implementation, we had participants leave the room after completing a final pre-survey before they would be able to discuss their roles/strategies with other players.

Practical Guidance and Lessons Learned

In the following section, we will provide a set of practical lessons that we developed based on difficulties that arose during various experimental sessions or related tasks (e.g., transferring data). While the applicability of each of these will depend on the personnel and experiment specifics you choose to implement; our hope is that these lessons will help you avoid a good deal of unproductive work in getting this experiment to work effectively.

Lesson 1: Minimize facilitator mental effort and decision-making While the provided facilitator script may seem simple to implement, it can require a lot of mental effort to ensure that all participants are engaging with the experiment appropriately. In many instances, participants will be confused by a part of the procedure; a piece of equipment will fail, an interruption will occur, or one of many other unanticipated issues may arise. In our experience, the number of tasks required for experiment implementation is challenging to be conducted by a single researcher. We usually used at least 3 researchers to manage all equipment, including data entry, working with camera equipment, troubleshooting technical issues, and interacting one-on-one with participants that required individual assistance. Because of the difficulty of this task, we recommend developing resources that minimize the effort required of researchers during the experiment by creating a narrow, specific script to follow. In our implementation, we carried separate versions of the experiment script, designed for each location, for each round, for each possible number of group size. By maintaining separate copies of the script, the facilitator does not have to determine the appropriate round information at the start of each round and only needs to read as they progress through the script. Similarly, we recommend practicing and recording steps required for the start of a session. If multiple researchers are involved, one can read the facilitator script to the participants, while another performs data entry and manages camera equipment to reduce time requirements. In addition to reducing effort, creating specific scripts will likely also decrease errors, increasing consistency, and improving the comparability of collected data.

Lesson 2: Use as much of the same equipment/infrastructure as possible between sites   While carrying equipment from site to site can be expensive and difficult, it is unrealistic to assume that sufficient on site-specific equipment will be available to create consistent experiment configurations. Table sizes, lighting, room size, power availability, internet connectivity, and noise level among a number of other site features will vary. A good deal of time and effort will be required to ensure data consistency. Bringing tables, networking equipment, computers, cameras, and power equipment with you will reduce this effort and increase consistency.

Lesson 3: Be prepared for lack of attendance   Attendance by participants that have signed up and/or completed a pre-survey will vary. In some instances, fewer than a third of participants registered for a session arrived on the day of the experiment. In an experiment that requires a group, this results in a canceled data collection session. We developed a few ways to avoid this issue:

(a) Maintain a waitlist of participants, informing them that they are on the waitlist and may only participate if fewer than the maximum number of participants arrive
(b) Send multiple reminders to participants through email and text message
(c) Ask those conducting on-site recruitment to engage with the sample population to emphasize the importance of attending their scheduled sessions

However, it is likely that some sessions will inevitably need to be canceled due to a lack of attendance despite these efforts. For these cases, you may be required to offer some form of compensation to participants though they were not able to complete the experiment. If another experiment that does not have the same group requirements is available, you may also consider redirecting them there.

Lesson 4: Use a temperature-controlled room or temperature-tolerant equipment if possible   Temperature can also impact your ability to collect data. First, and importantly, temperatures that are too hot or too cold can impact participants if they are uncomfortable. Changing levels of temperature due to lack of comfort likely adds an undesired confound to your data. For example, when participants were too cold, participants changed posture and shivered, changing the meaning of a variety of measurable body signals. Camera and computer equipment can also fail if it becomes too hot. In our implementation, our computer equipment would overheat in rooms with insufficient air conditioning, resulting in experiment slowdown and loss of video data.

Lesson 5: Prepare a manual backup for all experiment procedures that rely on technology   Cameras, computers, speakers, networking equipment, power equipment, lights, and any other form of technology can fail for a variety of unexpected reasons. It's possible that equipment will be lost in transit, dropped on accident, fail a software update, or be subject to a number of other possible issues. To prevent wasting participant and researcher time, having a plan for what to do when any particular piece of equipment fails can result in still collecting some data, even if not all the data from that session perfectly. We always carried at least 2 backup computers and had to swap out computers on multiple occasions at every data collection site.

Lesson 6: Backup all data as soon as possible after it is collected When dealing with large amounts of data, hard drives will inevitably fail. In addition to possibly losing data, insufficient storage can result in being unable to collect more data. By incorporating the creation of backup copies of all data that is collected into your experiment routine, you can dramatically reduce the likelihood that less than a single session-worth of data is lost. Using separate hard drives for backup means that if one hard drive fails, another can be used to continue to collect data while a replacement is obtained. In our implementation, we did have to replace hard drives using local sources for this reason. Maintaining backup copies of all data ensured that in spite of this, we were able to keep all of our collected data. Had the only copy of a particular piece of data resided on the hard drive that failed, we would have lost that data.

Lesson 7: Synchronize data sources as they are collected Though your research goal may vary, it is very likely that you will have multiple sources of data that you will want to synchronize. For example, if a participant completes a pre-survey at home, but completes the in-game survey on a different computer, you will need some way to synchronize those pieces of data. Similarly, you will likely want to synchronize audio/video files or facilitator recorded information (like game score) with survey data. In our experience, we cannot rely on participants to consistently enter uniquely identifying pieces of information. Even if provided specific codes for entering into forms, many will either forget or enter information incorrectly. Going back through the data after it has been collected can also be difficult, if not impossible, to resynchronize data. For this reason, we recommend that researchers enter all information required to link various sources of data themselves as it is being collected. For participant video files, name each file according to the unique identifier of the participant. For survey instruments, have the facilitator complete the first page that contains relevant identifying information.

Lesson 8: Coordinate for international human subjects research approval early Reviewing protocols for international human subjects research can require a substantial amount of time from researchers and those coordinating with target research sites. In some cases, several months may pass, ultimately to discover that data collection will not be possible at a potential target site. In our experience, planning data collection before being able to travel to an international site often took 6 months or more.

Lesson 9: Plan consistent interventions for instances where subjects are not interacting appropriately In many cases, particularly when participants were not familiar with similar games, participants were hesitant to interact during the experiment. Because participant interactions are an important part of the data to collect, we attempted to encourage less comfortable participants to participate. This, however, runs the risk of changing the nature of the game by guiding the discussion towards a particular outcome if care is not taken. For this reason, we incorporated into the facilitator script, specific neutral discussion-prompting dialog.

We recommend either adopting these specific phrases or conducting practice sessions to observe the type of interactivity that participants will have and create consistent prompting dialog to ensure consistent data collection.

Concluding Thoughts

We hope this description of our experimental procedures, tools, and reasoning has been helpful for understanding our implementation of this experiment or designing of a similar one. Much effort went into the creation and implementation of the experiment, ultimately resulting in a set of unique and valuable data. Many lessons have been learned along the way, and we hope that using this chapter as a guide, you will avoid some of the less productive aspects of designing and implementing an experiment such as this one.

Acknowledgement We are grateful to the Army Research Office for funding much of the work reported in this book under Grant W911NF-16-1-0342.

Funding Disclosure This research was sponsored by the Army Research Office and was accomplished under Grant Number W911NF-16-1-0342. The views and conclusions contained in this document are those of the authors and should not be interpreted as representing the official policies, either expressed or implied, of the Army Research Office or the U.S. Government. The U.S. Government is authorized to reproduce and distribute reprints for Government purposes notwithstanding any copyright notation herein.

References

Hofstede, G. (1983). The cultural relativity of organizational practices and theories. *Journal of International Business Studies, 14*(2), 75–89.

Singelis, T. M., Triandis, H. C., Bhawuk, D. P., & Gelfand, M. J. (1995). Horizontal and vertical dimensions of individualism and collectivism: A theoretical and measurement refinement. *Cross-Cultural Research, 29*(3), 240–275.

Zhou, L., Sung, Y., & Zhang, D. (2013). Deception performance in online group negotiation and decision making: The effects of deception experience and deception skill. *Group Decision and Negotiation, 22*(1), 153–172.

Chapter 5
Dominance in Groups: How Dyadic Power Theory Can Apply to Group Discussions

Norah E. Dunbar, Bradley Dorn, Mohemmad Hansia, Becky Ford, Matt Giles, Miriam Metzger, Judee K. Burgoon, Jay F. Nunamaker, and V. S. Subrahmanian

Dominance in Groups: How Dyadic Theories Can Apply to Group Discussions

Games are often used to train members of law enforcement, the military, and other practitioners to learn a variety of skills (e.g., Kapp 2012; Raybourn 2007; Miller et al. 2019). It is difficult to find real-world experiences for them to hone their skills, so games and simulations are an essential tool for laboratory experiments to examine how different types of personalities and cultures react to similar situations in a controlled environment. Therefore, in order to better understand how dominance plays a role in deception, and how groups of deceivers coordinate, a study was designed with groups of participants playing a face-to-face deception game, based on the party game variously known as *Mafia*, *The Resistance*, or *Werewolf*.

One objective of the game was to examine the role of dominance in how deception is enacted and detected. Data were collected from around the world, including in Israel, Singapore, Hong Kong, Fiji, Zambia, and three locations within the U.S. In what follows, we outline the ways in which groups differ from dyads in terms of the dominance strategies they use, review theories relevant to the study of dominance in dyads, and expand on these theories to make predictions at the group level when deception is involved. We then describe our game-based test of these hypotheses and research questions in a field experimental setting, and we discuss the implications of our findings for future research on deception in groups.

N. E. Dunbar (✉) · M. Hansia · B. Ford · M. Giles · M. Metzger
University of California, Santa Barbara, Santa Barbara, CA, USA
e-mail: ndunbar@ucsb.edu

B. Dorn · J. K. Burgoon · J. F. Nunamaker
University of Arizona, Tucson, AZ, USA

V. S. Subrahmanian
Dartmouth College, Hanover, NH, USA

© Springer Nature Switzerland AG 2021
V. S. Subrahmanian et al. (eds.), *Detecting Trust and Deception in Group Interaction*, Terrorism, Security, and Computation,
https://doi.org/10.1007/978-3-030-54383-9_5

Dominance and Deception

Dominance, as demonstrated through group members' verbal and nonverbal communication behaviors, is an effective strategy of social influence and power (Dunbar et al. 2016). Dominance is apparent through "control attempts" by an individual that are intended to change the behavior of the other interactant (Rollins and Bahr 1976). The role an individual has in the relationship between two people affects the amount and type of dominance displayed. By displaying more dominance, a deceiver attempts to appear confident, poised, credible and persuasive (Burgoon and Qin 2006). A recent review of nonverbal cues of dominance found that dominance can be expressed in a multitude of ways, including kinesics (using gaze, facial expressions, head pose, and gestures), the voice (such as pitch, speaking rate, pauses, and volume), and the use of personal space and the artifacts within the space (Dunbar and Bernhold 2019). Hall, Coats, and LeBeau (2005) conducted a meta-analysis in which they examined 27 different nonverbal cues of dominance in 120 different studies. They distinguished between "perceived" dominance, which is subjective and could be based on stereotypical beliefs, and "actual" dominance based on objectively-viewed displays by more dominant individuals. They emphasize the need to examine both perceived and actual dominance, suggesting that stereotypes and beliefs about dominance are stronger than actual differences between powerful and powerless people. We examine both perceived dominance and actual dominance behaviors in the overall project although the perceptions of dominance are the focus of this particular chapter.

The link between dominance and deception has been well documented (e.g. Dunbar et al. 2014, 2016; Zhou and Zhang 2006). Dominant members of groups are typically high-status individuals who have different motivations for deception and different deceptive strategies available to them compared to their low-status counterparts (Lindsey et al. 2011). Since dominance connotes credibility, panache, and interpersonal persuasiveness (Burgoon et al. 1998), it is often the case that liars who use dominant strategies are more likely to be believed.

Dyadic Power Theory

Power is an integral part of group dynamics (Dunbar et al. 2014). Dyadic power theory (DPT; Dunbar 2004), explains how the interpersonal power experienced by two interactants affects their communication strategies, including dominance displays. DPT is organized into pre-interactional factors that affect the relationship, the interactive process that occurs within the conversation, and the post-interactional implications of that process (Dunbar et al. 2014). Previous studies have investigated the use of dominance as a deception strategy, sometimes pairing DPT with a related theory, interpersonal deception theory (IDT; Burgoon and Buller 2008). For example, Dunbar et al. (2015) found that deception was a source of power because deceivers have access to information that the deception targets lack, notwithstanding that in certain relationships, the reverse is true by virtue of one's designated status (e.g., parent-child, officer-civilian, etc.).

DPT describes the nonlinear relationship between perceived power and control. It proposes that actors who perceive their relative power as extremely high or low compared to their partner will make fewer control attempts, while partners who perceive their relative power differences as small or moderate will make more control attempts. The rationale is that those in high-power do not need to exercise their power because their status symbols are sufficient to elicit deferent behavior from the other. Low-power others, by contrast, eschew making dominance attempts because they do not expect to succeed. It is those in the middle of the power continuum who feel freer to make dominance attempts to improve their power position in the relationship. These hypotheses have been supported in numerous studies examining a variety of relationship types. For example, in an experiment with dyads, Dunbar and Abra (2010) found that although high-power partners had structural power over their low-power counterparts, they preferred not to overrule them and veto their decisions. Instead, they preferred to engage their partner in a debate, giving the illusion of equality, but in the end, they exerted their power to make decisions. Interestingly, Dunbar and Johnson (2015) found that equal power partners were more likely to lie to one another than people in the high or low power relational position.

This study examined if a similar strategy is employed in groups. DPT would predict that less dominance should be used by group members with unequal power than those in which at least some group members have equal power. However, group tasks may require different communication strategies than those that work in dyads. Some alternatives: In keeping with DPT, high-power group members might give the illusion of inclusivity in decision-making so they can use dominance more subtly while less powerful members are being more vocal. However, high-power members might instead opt for being more assertive to gain more floor time and persuade multiple other group members. Low-power group members might decide they have nothing to lose by adopting a dominant strategy if they are losing influence and they perceive strength in their numbers. One of the goals of the current study was to identify dominance patterns and test the extent to which DPT can predict group-level behavior.

Group Behavior

Although scholars who study both groups and dyads are often interested in similar phenomena (e.g., persuasion, social influence, cooperation, and negotiation), groups are fundamentally different from dyads. For example, Moreland (2010) argues that individuals are likely to experience stronger and more negative emotions in dyads than in groups and they prefer small groups to large ones because individuals in dyads have a more direct experience with one another. Moreland further finds that dyads are simpler than groups because they involve only one relationship.

Marett and George (2004) develop the proposition further, noting that with multiple receivers, a suspicious receiver may not react at all, receivers may take different stances during the conversation or the same receiver may react at different times during the conversation. The task of monitoring all of this becomes exceedingly complex. Adding to this complexity is the fact that some influences operate at the individual level and others at the group level (Burgoon and Dunbar 2018). For example, individual personality differences may influence how a receiver regards sender behavior, whereas gender composition of the group is the same for all its members. Because of these differences between groups and dyads, it can be problematic to generalize dyadic-level theories to the group level. On the other hand, Williams (2010) argues that though groups larger than two might differ in some ways from those that have only two members, that does not mean that research using dyads cannot contribute to our understanding of larger groups.

The methods needed to study groups also differ from those which are employed to study dyads. Because of the interdependence that occurs among group members, it can be difficult to accurately distinguish between the group and dyadic levels (Giles et al. 2018). When dyadic communication occurs within a group setting, other members of the group are influenced by that communication, even when it is not addressed to them (Moreland 2010), and people often "talk to the room" rather than to just one individual. Talking to the room is a characteristic of high-functioning teams (Kolbe et al. 2014, p. 1254). Whether a comment or behavior is directed toward one group member or the group as a whole is an important distinguisher between dyadic and group communication.

In groups, as opposed to dyads, deceivers will likely use dominance strategies differently. Zhou and Zhang (2006) compared dominance in communication between deceivers in dyads and triads, arguing that deception should be more difficult in triads as there are more truth tellers to catch lies. The researchers found that deceivers exhibited more dominance in triads than in dyads, ostensibly because managing the communication of two others (as opposed to one) requires actively taking control of the situation.

When group members lie, they have knowledge that the other group members do not: they know they are lying. By knowing this additional piece of information, deceivers have power in the dyad or group (Dunbar et al. 2014). On the other hand, deceivers need to hide the fact that they have this additional piece of information, and therefore cannot disclose that they have additional power. In the context of a strategy game, too much dominance may lead to suspicion from other group members. Therefore, those who engage in deception need to balance two competing goals: acting in a dominant way in order to persuade others, while at the same time hiding the fact that they are being deceptive, which may require a more passive approach. This comports with Dunbar et al.'s (2014) prediction that deceivers will adopt one of two strategies: a defensive or "flight" strategy in which the deceiver becomes more submissive, or a more aggressive "fight" strategy. They might also dynamically alternate between these two strategies over time which makes predicting any one individual's behavior in the group very difficult.

While we agree with Moreland (2010) that many processes occur differently in groups of three or more from how they occur in dyads, we also agree with Williams (2010) that many similar processes occur in dyads and groups, such as social facilitation and loafing, cooperation, and leadership. Thus, we contend that theories of dyadic interaction, such as those that examine power and dominance, have much to contribute to our understanding of group dynamics.

Group communication scholars have frequently studied problem-solving in groups. Deception detection is a problem-solving task (Zhou et al. 2013). Furthermore, in the case where more than one group member is being deceptive, those who are part of the deceptive sub-group are trying to achieve influence. Dominance, as demonstrated through group members' verbal and nonverbal communication behaviors, is an effective strategy of social influence and power (Dunbar et al. 2016), and is a key variable in the study of deception detection (Dunbar et al. 2014; Zhou and Zhang 2006).

Hypotheses and Research Questions

The experiment we created investigated the effects of deception on relational communication, including dominance, during group interaction. Participants in the experiment played a variant of the *Mafia* game in which they were divided into Spies and Villagers. The goal of Villagers was to root out bad actors (Spies) through a series of missions. Spies had the goal of making missions fail, which required them to employ deception to keep their Spy identity secret. In addition, the Spies knew who the other Spies were in the game, giving them increased informational power compared to the Villagers. This context is different from most others in which DPT has been tested because, while the Spies have greater power, they also have an interest in keeping their power secret. They might want to behave dominantly to ensure mission failures, but they also needed to do so surreptitiously so that the Villagers did not suspect them to be Spies right away. In this experiment, because the Spies were engaged in a persuasive task, we predicted the following:

H1: Spies (deceivers) will be perceived as more dominant than Villagers.

Previous literature has characterized dominant behavior as more confident, poised, credible and persuasive (Burgoon and Qin 2006). If Spies are more dominant than Villagers, it can also be predicted that:

H2: Perceptions of dominance will be positively correlated with perceptions of trustworthiness.

However, past research has been operationalized with just one deceiver (e.g., Zhou and Zhang 2006; Zhou et al. 2013). Rather than one group member having a goal that opposes all other group members, more frequently, groups involve subgroups and cliques. In our study, multiple people were assigned to the deceiver role, and deceivers knew who the other deceivers were, creating the possibility of intragroup dynamics amongst deceivers. In other words, in settings where a deceiver has other members of their subgroup, they have multiple strategies that they can employ to accomplish their goal. Based on this, we asked the following questions:

RQ1: Do Spies coordinate their dominance with the other Spies?
RQ2: Are Spies more or less likely to change their dominance dependent on if they are winning or losing?

As previously highlighted, deceivers have power simply by knowing something that truth tellers do not. However, there are other ways a participant can maintain informational power. Those who have previous experience playing similar games are already familiar with some potential strategies (Zhou et al. 2013). In a dyad, if one person has task expertise and the other does not, DPT would predict that there should be few displays of dominance because the power is uneven and favoring those with expertise. Therefore, the following was predicted:

H3: Players with previous game experience will exhibit fewer dominance displays than those without such experience.

Deceivers are not the only actors involved in deception. IDT presents deception as a dynamic process in which receivers adapt their communication as their suspicions are aroused. However, people are not very good at deception detection in dyadic interactions (Bond and DePaulo 2006). Groups may be better suited for identifying deceivers because there are more people to catch inconsistencies and vocalize suspicions (Zhou and Zhang 2006). From this perspective, deception detection in groups is essentially a group problem-solving task (Zhou et al. 2013).

Though there may not be much research examining deception in groups, there is a long history of examining groups' abilities to solve problems. In fact, one of the reasons groups can be more complex than dyads is that there are more individuals who may have a diversity of backgrounds, skills, personalities, and experiences (e.g., Bowers et al. 2000; Marett and George 2004; Moreland 2010; Zhou et al. 2013). Burgoon and Dunbar (2018) identified three additional salient factors (culture, relational context, and gender), among others, that affect group dynamics and therefore should influence how dominance is enacted. In what follows, we review these group- and individual-level factors for their likely influence on dominance displays and deception detection during group interaction.

National Culture Culture plays an important role in what traits are considered trustworthy (Doney et al. 1998), which ultimately affects which communicative strategies a deceiver can use, including dominance, to be most successful. Hofstede's (1983, 2011) theory of cultural differences organizes cultures on four dimensions: power distance (or how much a society accepts inequality), uncertainty avoidance (how stressful the unknown future seems), individualism-collectivism (how people are organized as individuals or groups), and masculinity-femininity (how sharply the roles between men and women are divided). These cultural dimensions are then examined in a comparison of national identity. For example, in a study by Fernandez, Carlson, Stepina, and Nicholson (1997), the U.S. is considered to have low power distance but high individualism, whereas China is considered collectivistic and has a high power distance, suggesting that power is distributed more unevenly in Chinese culture than U.S. culture.

Based on these different cultural dimensions, dominant behaviors could be seen as trustworthy in one cultural context but suspicious in another. Research has found that in individualistic cultures, where personal advancement is prioritized above the group, people tend to approach trust in a calculative way, meaning they weigh the benefits of trusting someone against the costs of potentially being taken advantage of. Those from collectivist societies (those who prioritize the group over personal gain), on the other hand, are more likely to evaluate the trustworthiness of others based on the extent to which they adhere to normative behavior (Doney et al. 1998). Given that collectivistic cultures tend to have power hierarchies, there should be little need for dominance displays because it is already clear who has power. Deviating from communicative norms by acting dominant is likely to arouse suspicion in collectivistic cultures. Our experiment was conducted in six different locations around the world that vary on power distance and collectivism. This gives us a unique opportunity to investigate the cultural differences in dominance displays, perceptions of trustworthiness, and deception. Therefore, we asked:

RQ3: What is the effect of culture on perceptions of dominance and trust?

Gender Psychologists have long studied gender differences in personality, attitudes, and behavior. For example, research has found that women tend to be lower in dominance and assertiveness than men (see meta-analytic results in Feingold 1994), and this holds true across cultures (Costa et al. 2001). Furthermore, men are more likely to exhibit dominance in mixed-sex groups (Koch et al. 2010).

The influence gender has on dominance is evident from a very young age. Boys and girls as young as preschoolers clearly differ in how they play with their peers. Preschool boys engage in dominant behavior when interacting with classmates. When attempting to retrieve a toy, they are much more demanding and direct when compared to preschool girls (Serbin et al. 1982). Girls are much more likely to use persuasion and polite request as opposed to the threats and physical force that the boys use. Although differences are present between genders from an early age, research suggests that even when exhibiting identical behaviors, men and women are evaluated differently (Burgoon et al. 1986). Interruptions are often perceived as dominant behaviors and there are multiple types of interruptions that occur (Baker 1991). According to Anderson and Leaper (1998), intrusive interruptions occur when an individual abruptly talks over and takes over another's turn at talk. Conversely, cooperative overlaps as an interruption encourage the individual who is already talking. Intrusive interruptions are perceived as dominant. Research on gender stereotypes has shown that when comparing intrusive interruptions, men's interruptions were reported less often and seen as more dominant. Women were labeled as less dominant and their behavior as more unfair or unnecessary (Orcut and Harvey 1985). Furthermore, as there are expectations about the extent to which men and women are expected to display dominance (e.g., Burgoon et al. 1983), it is also possible that there will be gender differences in how trustworthy dominance is perceived to be. Considering previous research on men in mixed-sex groups, we predicted group games would replicate the same expected pattern that:

H4: Males will be perceived as more dominant than females.

Gender has also been studied for its impact on deception and deception detection. Whereas some research has found that men are worse than women at detecting deception (deTurck 2009; McCornack and Parks 1990; Zhou et al. 2013), most other research has failed to support this difference (Ekman and O'Sullivan 1991) or has given women a slight edge (Burgoon and Dillman 1995; Burgoon et al. 2006). However, the evidence seems to suggest that when it comes to behavioral displays, gender exerts main effect influences on deception displays rather than interacting with deception. The same main effects may emerge with dominance and gender. Of interest here is whether interaction effects emerge.

Method

Our version of the *Mafia* game (more detail below) was designed to examine how relational communication relates to deception and its detection in face-to-face group contexts. The first series of analyses emphasized the relational communication dimension of dominance. The same procedures were implemented at eight universities across six countries.

Participants

Participants ($N = 689$; $M_{age} = 22.38$, $SD_{age} = 3.70$) were primarily college students at public universities, augmented by some participants recruited from the general public. Data collection took place in Arizona (9 games; $n = 60$), California (11 games; $n = 78$), and Maryland (10 games; $n = 70$) in the U.S., and Israel (9 games; $n = 62$), Singapore (12 games; $n = 84$), Hong Kong (15 games; $n = 113$), Fiji (14 games; $n = 105$), and Zambia (15 games; $n = 117$) internationally. Participants were recruited via email and advertisements on public message boards. The sample was 53% female and was ethnically diverse (although this varied by location). Participants were required to be proficient English speakers.

Procedure

Participants were scheduled with an online scheduling system that assigned them to a game. The sessions ranged from five to eight participants. After signing up, participants were sent an email with a unique identifier number and a link to a pre-survey, which included consent forms, cultural measures, and demographic questions to be completed in advance. Upon arrival at the research site, participants were randomly assigned to one of eight computer stations that were arranged in a circle (See Figs. 5.1 and 5.2). Each station was equipped with an identical student desk and computer tablet that had a built-in webcam. Also in the room were a desk for the facilitator, who controlled game activity electronically, an overhead camera, and another long-view webcam. Participants were given instructions for the game and were informed that they would be filmed by several cameras (e.g., overhead camera, webcam, tablet cam).

After all participants were seated, the experiment facilitator explained the rules of the game. Participants first took part in an ice-breaker get-acquainted activity in which each player asked a question of the person opposite him or her. This activity provided baseline data and allowed participants to become familiarized with the environment and other players. After this ice-breaker activity, players rated each other on several scales (see Measures). Participants then engaged in up eight rounds of the game for an hour, whichever came first. The fewest rounds played was three; the most, eight. After the second, fourth, and sixth rounds, and at the end of the game, participants filled out several scales about their attitudes about the other players. Participants were paid for participating and could earn were small financial bonus incentives for performing well in the game.

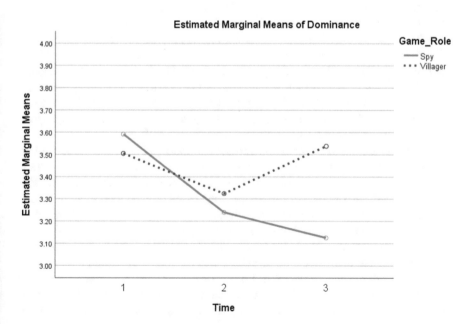

Fig. 5.1 Dominance ratings over time by game role

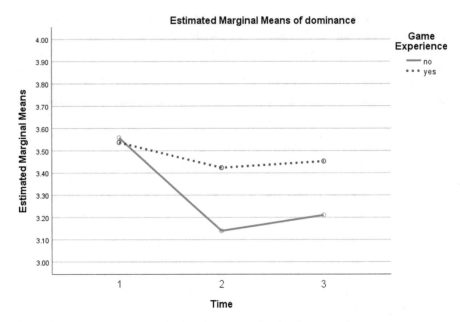

Fig. 5.2 Dominance ratings by all other players over time by player experience

Game Play

Similar to Zhou et al. (2013), we employed a version of the *Mafia* game, but one that more closely resembles the board game *The Resistance*. We pilot tested several versions of the game, and adapted the rules so as to best address the research questions. Players were randomly and secretly assigned to play deceivers (called "Spies"), or truth tellers (called "Villagers"). In games with five or six players, two were assigned to be Spies, and in games with seven or eight players, three were assigned to be Spies. The Spies were informed who the other Spies were, but the Villagers were unaware of other's roles. In order to thwart the Spies and win the game, Villagers had to depend on shared information to deduce the other players' identities.

The game consisted of a series of hypothetical "missions." First, at the beginning of each round, mission teams of varying size were to be formed. Second, players elected a leader. All players had to vote to approve or reject the leader. Third, the leader selected other players for these missions based on who they thought would help them win the round and, ultimately, the game. All players again voted, this time to approve or reject the leader's proposed team. Players voted secretly on their computer, and also voted publicly by raising their hands. They were allowed to vote differently on the computer than in public, and the facilitator would announce if there was a discrepancy in public and private votes, letting participants know that deception had occurred. Finally, those players who had been selected by the leader for the mission team voted in secret for the mission to succeed or fail. If the mission succeeded, Villagers received a point. If the mission failed, Spies received a point. Villagers thus won rounds by figuring out who the Spies were and excluding them from the mission teams. Spies won rounds by successfully being voted onto the mission teams and then causing missions to fail. The ultimate winner of the game (Spies or Villagers) was determined by which team won the most rounds. In addition to a monetary award for winning the game, individual players won monetary rewards by being voted as leader or by winning the game.

Dependent Measures

Game Outcome   In Zhou et al.'s (2013) Mafia study, deception detection success was operationalized as the truth tellers winning the game (i.e., if the truth tellers win, they must have accurately detected deception). Similarly, in this study, game outcome was a dichotomous variable measuring whether or not Spies ($n = 266$) or Villagers ($n = 423$) won the game. Since there were more Villagers than Spies overall, they had a greater opportunity to win the game, but the outcome of the games overall was nearly a 50/50 split on Spies and Villagers winning.

Trust   The extent to which participants trusted each of the other players was measured using a single-item repeated measure, which was asked after the ice-breaker, and then after every even-numbered round during the game. The item read: *Please rate how much you **trust** each player. Are they trustworthy or suspicious? A rating of 5 would mean they seem honest, reliable and truthful and 1 would mean you thought they were dishonest, unreliable and deceitful* (1 *Not at all* to 5 *Very* much; $M = 3.21$, $SD = 0.99$). Players rated all other players in their game after all even-numbered rounds. Because players had to rate one another three to five times during the game, we chose to use a single item to minimize fatigue. Since the spies were privy to objective information about their trustworthiness, our trust measures of interest were the ratings only from the Villagers, excluding Spies and self-ratings.

Dominance   The extent to which participants found other players to be dominant was measured using a single-item repeated measure (after the icebreaker and each of the even numbered rounds). Participants read the following text: *Please rate how **dominant** each player is. Are they active and forceful or passive and quiet? A rating of 5 would mean you thought they were assertive, active, talkative, and persuasive. A score of 1 would mean you thought they were unassertive, passive, quiet and not influential. Please mark any number from 1 to 5* (1 *Not at all* to 5 *Very* much; $M = 3.40$, $SD = 0.91$). Each player rated every other player on this item. As with the trust scores above, dominance scores for Spies included only dominance ratings from all the Villagers, excluding own dominance rating.

Previous Game Experience Participants' previous experience playing similar games was evaluated after the completion of the post-game measures. Participants indicated either that they had ($n = 308$) or had not ($n = 375$) played a similar game in the past (six participants left this item blank). Males were no more likely (45%) to have played the game before compared to females (44%), however, game experience was not evenly distributed across locations. While the majority of players answered "yes" to whether they had played Mafia or a similar game in Singapore (81%), Maryland (77%), California (63%), and close to half had game experience in Israel (43%), Hong Kong (50%), and Arizona (53%), the game was virtually unheard of in Fiji (14%) and Zambia (5%).

Results

To reduce the number of tests being conducted and to control for multicollinearity among dependent variables, unless otherwise stated, the hypotheses were tested with a single repeated measures MANOVA with "time" as the repeated measure. Dominance was measured at three time points: after the ice-breaker (Time 1), after the second round (Time 2), and after the last round played by a group (Time 3). The factors in the analysis were location (as a proxy for culture), the player's gender, the player's Spy or Villager role, whether they were on a winning or losing team, and whether the player had played Mafia (or a similar game) before. Due to the large number of factors, we limited the analysis to include only main effects and interactions with Spy/Villager role since that is the main variable of interest.

The dominance analysis showed that there was a significant main effect of Time, $F(1.91, 1273.61) = 35.04$, $p < .001$, $p\eta^2 = .05$. (Mauchly's Test of Sphericity indicated that the assumption of sphericity had been violated, therefore a Greenhouse-Geisser correction was used for the degrees of freedom). Pair-wise comparisons showed a significant decrease in dominance overall from Time 1 ($M = 3.56$, $SD = .68$) to Time 2 ($M = 3.30$, $SD = .95$) and Time 3 ($M = 3.38$, $SD = .94$).

H1 predicted that Spies would be perceived as more dominant than Villagers but we found the opposite was true. A significant Time X Game Role interaction $F(1.91, 1273.61) = 28.19$, $p < .001$, $p\eta^2 = .04$, indicated that spies were not significantly different from Villagers at Time 1 or Time 2, but at Time 3, Spies were perceived by fellow players to be less dominant ($M = 3.16$, $SD = .90$) than Villagers ($M = 3.52$, $SD = .94$); see Fig. 5.1. This was not moderated by whether the players were winning or losing the game, $F(1.91, 1273.61) = 1.55$, $p = .21$, $p\eta^2 = .002$. RQ2 asks whether Spies change their dominance over the course of the game depending on whether they are winning, but this analysis suggests that they do not. While Spies decreased their dominance over time, they did not change their strategy according to whether they were winning or losing.

H2 predicted that dominance would be positively associated with trustworthiness for Spies. We examined correlations between trust and dominance for all players at all three time-points and found positive correlations each time (Time 1 $r (688) = .23$, p < .01; Time 2 r (688) = .14, p < .01; Time 3 r (688) = .37, p < .01). But when examining ratings of Spies only (by Villagers), we found trust and dominance were significantly correlated only for Time 1 and Time 3 but not in Time 2 (Time 1 r (266) = .28, p < .01; Time 2 r (266) = .05, p = .42; Time 3 r (266) = .34, p < .01).

RQ1 asked how Spies managed their dominance in relation to their fellow Spies. To investigate this research question, we compared ratings of a given Spy with ratings of other Spies in the same group (made by Villagers). If deceivers are using similar levels of dominance, we would expect to see similar effects for all Spies in the group, but no effects specific to deceivers outside of this group-level effect. The group variable would capture this similarity or homogeneity within each group and any differences between groups would be attributable to Villager effects. If deceivers are using different levels of dominance, we would expect to see stronger effects on dominance ratings from the other Spies than from Villagers' dominance, after controlling for group effects. To investigate these effects, we used the following linear model:

$$Dominance_i = \beta_1 + \beta_2 OtherDominance_{ij} + \beta_3 Group_j + \varepsilon_{ij}$$

Where $Dominance_i$ is the average Villager-rated dominance of a particular Spy; $OtherDominance_{ij}$ is the average of Villager-rated dominance of other Spies in the same group; and $Group_j$ is a dummy variable for each particular group. Because group and other-deceiver ratings are confounded with groups of two or fewer deceivers (because there is only one other deceiver in the group), we only used observations from groups with three deceivers. Using this approach, we find that lower levels of other-Spy dominance results in higher dominance after controlling for group effects ($\beta_2 < 0$; $p < 0.01$), suggesting that deceivers use similar levels of dominance to each other. In other words, the behavior of the other two spies has a greater effect on the dominance of any given spy than the behavior of the villagers. The results of our linear model are presented in Table 5.1.

H3 investigated the role of previous game experience, with the assumption that those who have Mafia game experience will know to keep their identities hidden through lower dominance levels. A significant Time X Experience interaction, $F(1.91, 1273.61) = 8.65$, $p < .001$, $p\eta^2 = .013$, suggests the opposite is true. As Fig. 5.2 reveals, players with game experience did not differ from those without during the ice-breaker (Time 1), but during game play they were more dominant than players without game experience at Time 2 and Time 3. Game experience apparently was an informational resource that promoted dominance. Novice players also may have exhibited uncertainty about the rules of the game and thus appeared less dominant. This was not moderated by their role as a Spy or Villager in the game, $F(1.91, 1273.61) = 1.17$ $p = .30$, $p\eta^2 = .002$. The very small effect size, however, warrants viewing this as a very modest effect.

We also examined whether players with previous game experience would be trusted less than players without, given that knowing how to play the Mafia game might be perceived by other players as an unfair advantage. This did not prove to be true. A similar repeated measures MANOVA for trust as the one described above

Table 5.1 Effects of Other Team Members' Dominance on Spy Dominance

Variables	B	SE	ß
Dominance of other Spies	0.19	0.09	0.25*
Dominance of Villagers	0.00	0.12	0.00
Model Summary			
	df		2230
	R²		0.02
	F		2.52†

†$p < .10$; *$p < .05$

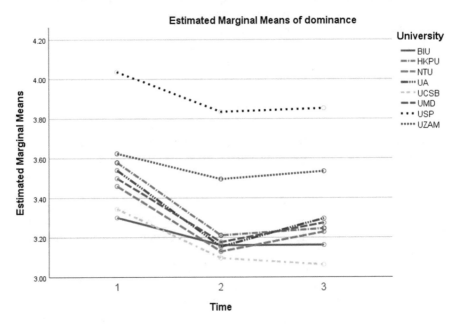

Fig. 5.3 Dominance ratings over time by university location

failed to produce a significant Time X Experience interaction, $F(1.18, 1212.80) = 1.53$, $p = .22$, $p\eta^2 = .002$. Players with experience in the game did not suffer from a trust deficit.

RQ3 questioned whether the culture of the location of the data collection would affect the dominance and trust ratings of other players. It will be recalled that there were eight locations for the experiment, including three in the U.S. and five in international locations. The Time X University interaction was not significant for dominance $F(13.35, 1273.61) = .68$, $p = .79$, $p\eta^2 = .007$. However, post-hoc comparisons showed that one location, the University of South Pacific (USP) in Fiji, had higher dominance ratings than any of the other locations and differed significantly from every other location ($p < .05$) at all three time periods (see Fig. 5.3). Fijian students also rated each other higher on the trust measure than the other locations, at least at Time 1 and 2 (see Fig. 5.4). In addition, on the dominance

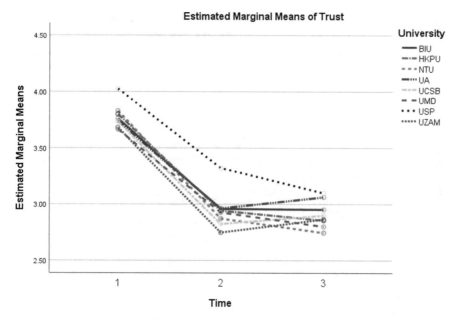

Fig. 5.4 Trust ratings over time by university location

measure, there was a significant 3-way interaction, Time X Location X Game Role $F(13.34, 1273.61) = 1.87$, $p = .028$, $p\eta^2 = .019$. It appears that players at different locations used different dominance strategies according to their role. For example, while the Fiji players were perceived to be the most dominant, this was especially true when they were in the Villager role (compare Fig. 5.5a, b). In Zambia, however, it was Spies who were more likely to adopt a dominant role, according to their peers' perceptions.

H4 predicted that males will be perceived as more dominant than females. The repeated measures MANCOVA described above did reveal a significant Time X Sex interaction, $F(1.90, 1273.61) = 4.40$, $p = .012$, $p\eta^2 = .007$. Males were seen as more dominant than females at all three time periods (Fig. 5.6). This was further amplified when women were in the Spy role. The Time X Sex X Game role interaction, and depicted below in Fig. 5.7a, b, was significant, $F(1.91, 1315.40) = 3.42$, $p = .034$, $p\eta^2 = .005$. However, the perception of dominant males did not have a positive result on the perceived trustworthiness of males, as gender was not a significant predictor of trust. The Time X Sex interaction, $F(1.88, 1254.22) = 2.17$, $p = .12$, $p\eta^2 = .003$, was not significant.

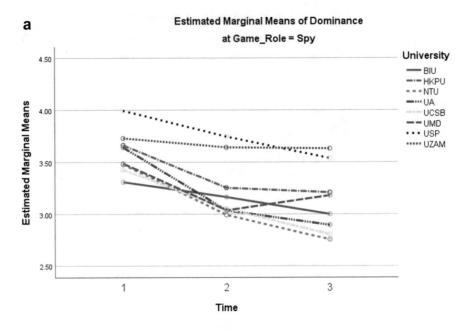

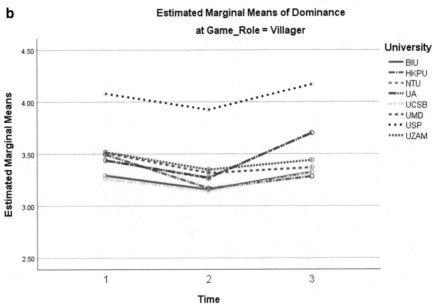

Fig. 5.5 (**a**) Dominance ratings over time by location for Spies. (**b**). Dominance ratings over time by location for Villagers

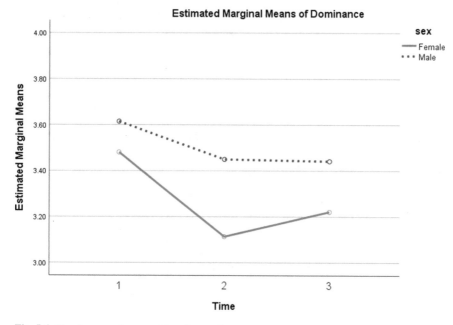

Fig. 5.6 Dominance ratings over time by gender

Discussion

These results have numerous interesting implications. They reveal that although dominance is enacted differently in groups than it is in dyads, there are some important similarities between the two situations. While DPT predicts that equal power dyads are the most dominant compared to their unequal power counterparts, in typical studies of DPT, the power relationship is known to all the participants. In this context, Spies had information that the Villagers lacked, giving them what Raven, Centers, and Rodrigues (1975) would call *informational power*. In addition, some players had experience with the game before, which gave them both knowledge and confidence that can be perceived as dominant. Thus, Spies had to choose between a confrontational (fight) approach to persuade the Villagers to trust them or a more submissive (flight) approach to hide their identity and "fly under the radar." We predicted that spies would adopt a more dominant style because that would ensure their place on mission teams to give them an opportunity to fail the missions. However, Spies were apparently more concerned with being discovered and instead opted for a more submissive style, although this strategy did not reveal itself until closer to the end of the game.

Villagers' ratings of Spies on dominance decreased over time, whereas it increased for other Villagers as they came to their final game round. This is consistent with DPT's prediction that those with equal power (Villagers rating other Villagers, for example) would be more dominant than those with unequal power (in

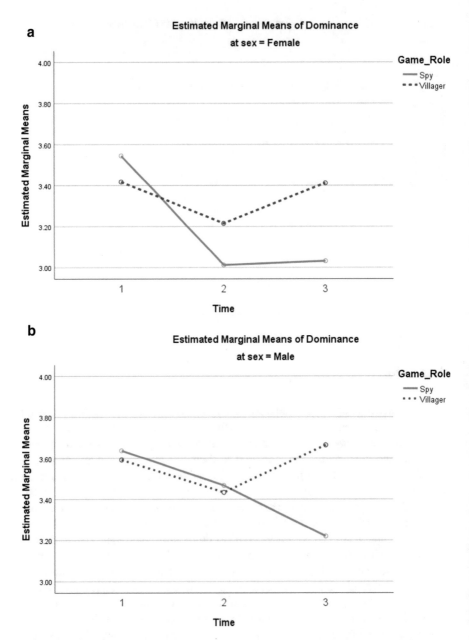

Fig. 5.7 (a) Dominance ratings over Time by Game Role for Females. (b). Dominance ratings over Time by Game Role for Males

the case of Villagers rating Spies). However, the reason that Spies were perceived as less dominant cannot be explained thoroughly by DPT. Here, the Spies were less dominant because they took the "flight" strategy, while people with high power status are typically less dominant because they pretend to be inclusive when making decisions by DPT. These results suggest that while some aspects of DPT might be used to make dominance predictions in group contexts, other parts of the theory need to be adjusted for specific deception contexts. Broadening DPT to account for a wider range of reasons for why people with higher (informational) power might display lower dominance. It may be premature to declare that DPT can be used to make dominance predictions in group contexts but this first test should encourage future tests.

The group context also introduces interesting confounding variables which are not present in dyads. Our research indicates that deceivers coordinate their dominance relative to each other, which introduces gradation into an interaction that is more easily considered in binary terms for dyads (high and low dominance). When two deceivers enact different dominance strategies while simultaneously engaging in deceptive behavior, the relationship between dominance and trustworthiness cannot be described in terms of a simple one-way relationship. It would have been beneficial for Spies to adopt different strategies so that one Spy could take the pressure off the others, but they did not coordinate their behavior in this way. Instead, we saw a general decline in dominance for Spies across the board.

In addition, although dominance was correlated with trustworthiness, the increased dominance on the part of Villagers near the end of the game did not give them an edge in winning the game. Dominant players may have been more confident in their abilities or perhaps had more charisma and extroverted styles that led them to appear more trustworthy. Despite their increased dominance, Villagers were not more likely to win (the win ratio was nearly 50/50 for Villagers and Spies). Players with previous experience were also perceived to be more dominant but it also did not change the perceptions of their trustworthiness. Thus, the link between dominance and trustworthiness does not seem to result in outcomes that are beneficial to the players.

The national cultural context for the games also provided interesting information in the way that dominance is enacted. Two countries were perceived to be using higher levels of dominance than the others: Fiji and Zambia. Our local host in Fiji (J. Johnson, personal communication, January 14, 2019) observed that the Mafia game is much more confrontational than the games typically played in Fiji. Because players must accuse one another of wrong-doing, this would be perceived as much more dominant than what would be normally expected in a social gathering. Similarly, our local host from Zambia (R. Muchangwe, personal communication, June 10, 2019) felt that Zambians might be reluctant to challenge others even when they perceive wrongdoing and so this context was also somewhat unfamiliar to them. Indeed, players had much less experience with a Mafia-style game in both of those locations, so the unfamiliarity with the game may have also prompted the higher dominance ratings. Interestingly, those high dominance ratings translated into higher trust ratings in Fiji but not in Zambia, which ranked second highest on dominance but lowest on perceptions of trust (see Figs. 5.3 and 5.4). Zambia also saw their Spies as more dominant than Villagers, bucking the overall trend of Villager dominance. This variability in responses by culture indicates that predictions about dominance must be calibrated by location, not treated as a generic pattern.

Gender also played an interesting role in the perceptions of dominance. Females were no more likely to be inexperienced with the game than males, but males were far more likely to use a dominant strategy overall. This effect was further magnified by game roles. Female Spies were rated similarly in dominance to their male counterparts immediately after the icebreaker (before roles were assigned and people were speaking about themselves truthfully). But as soon as the game began and spy roles were assigned, female Spies immediately dropped in dominance (see the Time 2 ratings in Fig. 5.7a, b).

Limitations

While provocative, the results of this study must be interpreted with certain limitations in mind. Specifically, the samples we used were not representative of the populations in each location we studied. College students comprised the game players (both Spies and Villagers) in all eight locations. Although there is no theoretical reason to expect younger players to act differently than older players in the context of our game, we cannot rule out the possibility that results may be different if a broader range of participants from each location is included. We must also take care not to generalize beyond our data. Although this study is among the most ambitious to examine dominance, trust, and deception detection cross-culturally, participants were limited to English speakers only, even in foreign countries. Moreover, participants in some of the countries (e.g., Israel) were speaking English as a second language, while in others English was all players' first or at least official language (e.g., U.S.).

Another limitation of this research, at least in terms of generalizing the results to real-world deception situations, is the fact that we used a game to stimulate deception. Of course this means that participants were aware that they were in a contrived rather than a true decision-making situation, and although the stakes for successful deception and deception detection abilities were real monetary rewards, this is quite different than the stakes that would be involved in most group deception settings in the real world. Further research in more naturalistic settings is thus needed to understand if the results of our study hold up in other contexts.

Future Directions

Future examinations from this dataset and others should examine not only the perceived dominance but the actual dominance behaviors evident through kinesics, vocalics, linguistic choices, and other objective cues. As mentioned earlier, we collected extensive video of the games and our next steps include analyses of actual dominance displays during the game by both Spies and Villagers.

Acknowledgement   We are grateful to the Army Research Office for funding much of the work reported in this book under Grant W911NF-16-1-0342.

Funding Disclosure   This research was sponsored by the Army Research Office and was accomplished under Grant Number W911NF-16-1-0342. The views and conclusions contained in this document are those of the authors and should not be interpreted as representing the official policies, either expressed or implied, of the Army Research Office or the U.S. Government. The U.S. Government is authorized to reproduce and distribute reprints for Government purposes notwithstanding any copyright notation herein.

References

Anderson, K. J., & Leaper, C. (1998). Meta-analyses of gender effects on conversational interruption: Who, what, when, where, and how. *Sex Roles, 39*(3–4), 225–252. https://doi.org/1018802521676.

Baker, M. A. (1991). Gender and verbal communication in professional settings: A review of research. *Management Communication Quarterly, 5*(1), 36–63. https://doi.org/10.1177/0893318991005001003.

Bond, C. F., & DePaulo, B. M. (2006). Accuracy of deception judgments. *Personality and Social Psychology Review, 10*, 214–234. https://doi.org/10.1207/s15327957pspr1003_2.

Bowers, C. A., Pharmer, J. A., & Salas, E. (2000). When member homogeneity is needed in work teams. *Small Group Research, 31*, 305–327. https://doi.org/10.1177/104649640003100303.

Burgoon, J. K., & Buller, D. B. (2008). Interpersonal deception theory. In L. A. Baxter & D. O. Braithwaite (Eds.), *Engaging theories in interpersonal communication: Multiple perspectives* (pp. 227–239). SAGE. https://doi.org/0.1111/j.1468-2885.1996.tb00127.

Burgoon, J. K., & Dillman, L. (1995). Gender, immediacy and nonverbal communication. In P. J. Kalbfleisch & M. J. Cody (Eds.), *Gender, power, and communication in human relationships* (pp. 63–81). Hillsdale: Erlbaum.

Burgoon, J., & Dunbar, N. (2018). Coding nonverbal behavior. In *The Cambridge handbook of group interaction analysis. Cambridge handbooks in psychology* (pp. 104–120).

Burgoon, J. K., & Qin, T. (2006). The dynamic nature of deceptive verbal communication. *Journal of Language and Social Psychology, 25*(1), 76–96. https://doi.org/10.1177/0261927x05284482.

Burgoon, M., Dillard, J. P., & Doran, N. E. (1983). Effects of violations of expectations of males and females. *Human Communication Research, 10*, 283–294.

Burgoon, J. K., Coker, D. A., & Coker, R. A. (1986). Communicative effects of gaze behavior: A test of two contrasting explanations. *Human Communication Research, 12*(4), 495–524. https://doi.org/10.1111/j.1468-2958.1986.tb00089.x.

Burgoon, J. K., Johnson, M. L., & Koch, P. T. (1998). The nature and measurement of interpersonal dominance. *Communications Monographs, 65*(4), 308–335. https://doi.org/10.1080/03637759809376456.

Burgoon, J. K., Buller, D. B., Blair, J. P., & Tilley, P. (2006). Sex differences in presenting and detecting deceptive messages. In D. Canary & K. Dindia (Eds.), *Sex differences and similarities in communication* (2nd ed., pp. 263–280). Mahwah: LEA.

Costa, P. T., Terracciano, A., & McCrae, R. R. (2001). Gender differences in personality traits across cultures: Robust and surprising findings. *Journal of Personality and Social Psychology, 81*, 322–331. https://doi.org/10.1037//0022-3514.81.2.322.

deTurck, M. A. (2009). Training observers to detect spontaneous deception: Effects of gender. *Communication Reports, 4*, 81–89. https://doi.org/10.1080/08934219109367528.

Doney, P. M., Cannon, J. P., & Mullen, M. R. (1998). Understanding the influence of national culture on the development of trust. *Academy of Management Review, 23*, 601–620. https://doi.org/10.5465/amr.1998.926629.

Dunbar, N. E. (2004). Dyadic Power theory: Constructing a communication-based theory of relational power. *Journal of Family Communication, 4*, 235–248. https://doi.org/10.1080/15267431.9670133.

Dunbar, N. E., & Abra, G. (2010). Observations of dyadic power in interpersonal interaction. *Communication Monographs, 77*, 657–684. https://doi.org/10.1080/03637751.2010.520018.

Dunbar, N. E., & Johnson, A. J. (2015). A test of dyadic power theory: control attempts recalled from interpersonal interactions with romantic partners, family members, and friends. *Journal of Argumentation in Context, 4*(1), 42–62. https://doi.org/10.1075/jaic.4.1.03dun.

Dunbar, N. E., & Bernhold, Q. (2019). Interpersonal power and nonverbal communication. In C. R. Agnew & J. J. Harmon (Eds.), *Power in close relationships* (pp. 261–278). Cambridge, UK: Cambridge University Press.

Dunbar, N. E., Jensen, M. L., Bessarbova, E., Burgoon, J. K., Bernard, D. R., Harrison, K. J., & Eckstein, J. M. (2014). Empowered by persuasive deception. *Communication Research, 41*, 852–876. https://doi.org/10.1177/0093650212447099.

Dunbar, N. E., Jensen, M. L., Burgoon, J. K., Kelley, K. M., Harrison, K. J., Adame, B. J., & Bernard, D. R. (2015). Effects of veracity, modality, and sanctioning on credibility assessment during mediated and unmediated interviews. *Communication Research, 42*, 649–674. https://doi.org/10.1177/0093650213480175.

Dunbar, N. E., Gangi, K., Coveleski, S., Adams, A., Bernhold, Q., & Giles, H. (2016). When is it acceptable to lie? Interpersonal and intergroup perspectives on deception. *Communication Studies, 67*(2), 129–146. https://doi.org/10.1080/10510974.2016.1146911.

Ekman, P., & O'Sullivan, M. (1991). Who can catch a liar? *American Psychologist, 46*, 913–920. https://doi.org/10.1037/0003-066X.46.9.913.

Feingold, A. (1994). Gender differences in personality: A meta-analysis. *Psychological Bulletin, 116*(3), 429. https://doi.org/10.1037/0033-2909.116.3.429.

Fernandez, D. R., Carlson, D. S., Stepina, L. P., & Nicholson, J. D. (1997). Hofstede's country classification 25 years later. *The Journal of Social Psychology, 137*(1), 43–54. https://doi.org/10.1080/00224549709595412.

Giles, M., Pines, R., Giles, H., & Gardikiotis, A. (2018). Toward a communication model of intergroup interdependence. *Atlantic Journal of Communication, 26*, 122–130. https://doi.org/10.1080/15456870.2018.1432222.

Hall, J. A., Coats, E. J., & LeBeau, L. S. (2005). Nonverbal behavior and the vertical dimension of social relations: A meta-analysis. *Psychological Bulletin, 131*(6), 898. https://doi.org/10.1037/0033-2909.131.6.898.

Hofstede, G. (1983). The cultural relativity of organizational practices and theories. *Journal of International Business Studies, 14*, 75–89. https://doi.org/10.1057/palgrave.jibs.8490867.

Hofstede, G. (2011). Dimensionalizing cultures: The Hofstede model in context. *Online Readings in Psychology and Culture, 2*, 1–26. https://doi.org/10.9707/2307-0919.1014.

Kapp, K. M. (2012). *The gamification of learning and instruction*. San Francisco: Wiley.

Koch, S. C., Baehne, C. G., Kruse, L., Simmermann, F., & Zumbach, J. (2010). Visual dominance and visual egalitarianism: Individual and group-level influences of sex and status in group interactions. *Journal of Nonverbal Behavior, 34*, 137–153. https://doi.org/10.1007/s10919-010-0088-8.

Kolbe, M., Grote, G., Waller, M. J., Wacker, J., Grande, B., Burtscher, M., & Spahn, D. (2014). Monitoring and talking to the room: Autochthonous coordination patterns in team interaction and performance. *Journal of Applied Psychology, 99*, 1254–1267. https://doi.org/10.1037/a0037877.

Lindsey, L. L. M., Dunbar, N. E., & Russell, J. C. (2011). Risky business or managed event? Perceptions of power and deception in the workplace. *Journal of Organizational Culture, Communications and Conflict, 15*(1), 55. https://doi.org/10945/56702.

Marett, L. K., & George, J. F. (2004). Deception in the case of one sender and multiple receivers. *Group Decision and Negotiation, 13*, 29–44. https://doi.org/10.1023/B:GRUP.0000011943.73672.9b.

McCornack, S. A., & Parks, M. R. (1990). What women know that men don't: Sex differences in determining the truth behind deceptive messages. *Journal of Social and Personal Relationships, 7*(1), 107–118. https://doi.org/10.1177/0265407590071006.

Miller, C. H., Dunbar, N. E., Jensen, M. L., Massey, Z., Lee, Y-. H., Nicholls, S. B., Anderson, C., Adams, A. S., Elizondo Cecena, F. J., Thompson, W., & Wilson, S. N. (2019). Training law enforcement officers to identify reliable deception cues with a serious digital game. *International Journal of Game-Based Learning, 9*(3), 1–22. https://doi.org/10.4018/IJGBL.2019070101.

Moreland, R. L. (2010). Are dyads really groups? *Small Group Research, 41*, 251–267. https://doi.org/10.1177/1046496409358618.

Orcutt, J. D., & Harvey, L. K. (1985). Deviance, rule-breaking and male dominance in conversation. *Symbolic Interaction, 8*(1), 15–32. https://doi.org/10.1525/si.1985.8.1.15.

Raybourn, E. M. (2007). Applying simulation experience design methods to creating serious game-based adaptive training systems. *Interacting with Computers, 19*(2), 206–214. https://doi.org/10.1016/j.intcom.2006.08.001.

Raven, B. H., Centers, R., & Rodrigues, A. (1975). The bases of conjugal power. In R. E. Cromwell & D. H. Olson (Eds.), Power in families (pp. 217–232). Beverly Hills, CA: Sage.

Rollins, B. C., & Bahr, S. J. (1976). A theory of power relationships in marriage. *Journal of Marriage and the Family, 38*, 619–627. https://doi.org/10.2307/350682.

Serbin, L. A., Sprafkin, C., Elman, M., & Doyle, A. B. (1982). The early development of sex-differentiated patterns of social influence. *Canadian Journal of Behavioural Science/Revue canadienne des sciences du comportement, 14*(4), 350. https://doi.org/10.1037/h0081269.

Williams, K. D. (2010). Dyads can be groups (and often are). *Small Group Research, 41*, 268–274. https://doi.org/10.1177/1046496409358619.

Zhou, L., & Zhang, D. (2006). A comparison of deception behavior in dyad and triadic group decision making in synchronous computer mediated communication. *Small Group Research, 37*, 140–164. https://doi.org/10.1177/1046496405285125.

Zhou, L., Zhang, D., & Sung, Y. (2013). The effects of group factors on deception detection performance. *Small Group Research, 44*, 272–297. https://doi.org/10.1177/1046496413484178.

Chapter 6
Behavioral Indicators of Dominance in an Adversarial Group Negotiation Game

Steven J. Pentland, Lee Spitzley, Xunyu Chen, Xinran (Rebecca) Wang, Judee K. Burgoon, and Jay F. Nunamaker

Introduction

Negotiation involves executing strategies and tactics to satisfy conflicting interests of disparate parties. High self-interests and low concern for opposing interests encourages contentious strategies and tactics such as threats, positional commitments, and persuasive arguments (Pruitt 1983). Within the sphere of contentious strategies are dominance (Linell et al. 1988) and deception (Schweitzer et al. 2005), where individuals attempt to control the discussion and conceal/misrepresent information in favor of their interests. Here, we refer to negotiations involving *high-self-interests* and *low-other-interests* and employing dominant or deceptive strategies as *adversarial negotiations*.

In this chapter, we focus on *what* is said and *how* it is said during adversarial group negotiations. This research seeks to better understand characteristics of language and voice related to dominance and deception by analyzing linguistic content and paralinguistic features. These characteristics are traditionally studied with respect to dyadic interactions. However, groups are fundamentally different from dyads, experiencing less emotion, but involving more complex relationships than dyads (Moreland 2010). Further, hidden agendas during negotiations increase the tension felt by participants (Giordano et al. 2007) and deception increases emotional arousal and cognitive load.

S. J. Pentland
Boise State University, Boise, ID, USA
e-mail: stevenpentland@boisestate.edu

L. Spitzley (✉)
University at Albany, Albany, NY, USA
e-mail: lspitzley@albany.edu

X. Chen · X. (Rebecca) Wang · J. K. Burgoon · J. F. Nunamaker
University of Arizona, Tucson, AZ, USA

© Springer Nature Switzerland AG 2021   99
V. S. Subrahmanian et al. (eds.), *Detecting Trust and Deception in Group Interaction*, Terrorism, Security, and Computation,
https://doi.org/10.1007/978-3-030-54383-9_6

The complexity of adversarial group negotiations likely causes the verbal and nonverbal behaviors of group members to differ from behaviors found in similar dyadic interactions. This research investigates two questions: 1) What linguistic and vocalic features signal dominance and deception during adversarial group negotiations? 2) Does the role of a participant within an adversarial group influence his/her linguistic and vocalic behaviors? We begin this chapter by reviewing adversarial relationships, group versus non-group communication, and linguistic and vocalic cues of dominance and deception. We then outline the data extraction methodology used to generate voice and language features. Finally, we present and discuss our results.

Literature Review

Adversarial Relations

Though some negotiations focus on collaboration that fosters trust, traditional negotiations and bargaining are embedded in adversarial settings (Kelleher 2000). Adversarial relations are, "…relations that cause emotional distress, anger, or indifference" (Yang and Tang 2003, p. 97). These types of relationships can emerge through incongruent objectives and lack of trust (Oade 2011). The label *adversarial collaboration* (AC) describes group settings where members have competing interests but must reach a mutually agreed upon outcome. Cohen et al. (2000) describe AC as consisting of *secrecy* (concealment of information), *advocacy* (advancing individual interests), and *discovery* (strategic information disclosure). That is, aspiring towards a desired outcome amid competing interests is strategic in nature and may involve persuasion and deception. A natural aspect of any group is to work through individual differences in order to reach a common goal, and in some cases, opposition within a group can lead to better performance outcomes (Baldwin et al. 1997). However, more severe problems may emerge when opposition members choose to remain hidden, instead attempting to conceal their true intentions and thwart any and all progress the group is attempting to make. Migdal (2010) found that even a fraction of opposition members with hidden objectives can drastically decrease the odds of a favorable outcome for the majority. In computer-mediated communication, deception also has a negative organizational impact and causes employees to have higher levels of tension and lower levels of long-term effectiveness (Giordano et al. 2007).

Group members who adopt a dominant strategy may be perceived as more credible, persuasive (Burgoon et al. 1998) and likeable (Holtgraves and Lasky 1999), leading to a higher likelihood of achieving their objective. Dominance is displayed nonverbally in a variety of different ways (Hall et al. 2005). However, the type of behaviors displayed may differ based on sender-receiver variables such as gender, culture, age, personality, or any number of environmental circumstances. For instance, Chap. 5, Patterns of Dominance during a Game of Deception, hypothe-sized that deceivers (opposition members) will be more dominant compared to truth-tellers (majority) in adversarial settings, but found the opposite. Truthful group members perceived deceivers as less dominant and decreasingly dominant as time progressed. This finding suggests that a clandestine opposition may adopt a submissive strategy that more subtly thwarts group progress, since a dominant deception strategy may be viewed as too assertive. Because opposition members and majority members differed in their levels of dominance during our data collec-tion, an exploration of behavioral predictors of dominance and deception in groups may distinguish between members from each faction. Practitioners can apply these findings to guide group outcomes and identify opposition in settings like business negotiations, law enforcement, and fraud investigations.

Group Versus Non-Group Communication

Much of the prior literature on dominance and deception is based on dyadic interac-tions, and it is important to discuss the similarities and differences between dyads and groups. While Williams (2010) argues that similar processes occur whether teams are constituted of two or more people, dyads and larger groups differ in vari-ous ways. Compared with dyads, groups form and dissolve more slowly, and group members are less emotionally connected (Moreland 2010). Also, group size is an important factor in research of group behaviors (Thomas and Fink 1963; Vernham et al. 2016) and group support systems (DeSanctis and Gallupe 1987; Gallupe et al. 1992). Groups of various sizes differ in their ability to handle four functional prob-lems, namely sharing identities, generating resources, establishing norms, and exer-cising controls (Hare 1981). In addition, problems of production blocking and evaluation apprehension prevent large groups from generating more ideas than small groups (Gallupe et al. 1992).

The unique characteristics of groups make deception more difficult (Marett and George 2004) and deceivers may adopt a different strategy in a group environment. First, deceivers usually have to face majority influence in a group, and the presence of others and confronting other group members leads to higher arousal and more behavioral leakage (McCullagh and Landers 1976; Zajonc 1965). Second, group output is large and diverse in terms of the knowledge and perspectives brought by turn-taking and interactional patterns, so deceivers are likely to experience a higher cognitive load when monitoring group members and interpreting the suspicion lev-els of others (Zhou and Zhang 2006). However, larger groups also provide more opportunities for free-riding (Valacich et al. 1992). In groups of three to eight mem-bers, one or two individuals usually do most of the talking (Báles et al. 1951), so deceivers can comfortably "fly under the radar" using discretionary time to craft strategies. For instance, deceivers can passively observe and accommodate the behaviors of other group members (Giles and Baker 2008), and thereby better obscure their nefarious motives. However, large groups require more control and coordination from a leader (Stogdill 1974), making it more cognitively demanding for deceivers to exert influence when the situations call for active persuasion. Lastly, emotion-based cues to deception may be muted, since groups are generally less emotionally charged than dyads (Moreland 2010).

Groups also influence perceptions and relationships. Groups of larger sizes have less intimate information exchanges and lower levels of affective ties (Cartwright and Zander 1968; DeSanctis and Gallupe 1987), so the increased perceived distance between members of larger groups makes them more cautious to new messages and less tolerant of different views (Zhou and Zhang 2006). If the physical space is crowded for groups, people are more negative to each other (Hare 1981).

Linguistics

Written or spoken language communicates direct messages and subtle information about the sender. Linguistic content analysis can reveal attributes such as emotions, status, social standing, and cognitive processes (Tausczik and Pennebaker 2010). For instance, the use of more negative and self-reflective language correlates with depressed mental status (Rude et al. 2004). Linguistic analysis has received growing attention over the past 20 years as web-based text content balloons and more sophisticated text analysis techniques emerge. Sentiment analysis, part-of-speech tagging, and word tagging using lexicons are all mechanisms commonly used to understand the subtle characteristics of language. One of the goals of this paper is to better understand the relationships between language, dominance, and deception.

Dominance Language

Individuals who have a dominant status in a dyad or group may reveal their position through language. Linguistic cues of dominance include speech quantity, dominant phrases, subjunctive phrases, uncertainty, and emotion. Monopolizing behaviors, such as talking often and talking for longer duration, are a factor of dominance as measured by Norton (1983), and quantitative dominance, along with sequential and participatory dominance, is an important dimension of dominance in conversations (Itakura 2001). The sum of turns-at-talk and the amount of text indicates how much a speaker contributes to a conversation (Zhou et al. 2004b), and dominant people use more words (Itakura 2001). Moreover, dominant individuals are pictured as being assertive, influential (Burgoon et al. 1998), authoritative (Zhou et al. 2004b) and self-confident (Weisfeld and Linkey 1985). Correspondingly, they tend to use less hedging and display fewer hesitations, because hedging and hesitations lower evaluations of authoritativeness (Hosman 1989). Being influential and self-confident implies a more definitive speech style and less use of subjunctive language (Zhou et al. 2004b); phrases with certain modal verbs such as "could" and "might" are replaced with ones of greater certainty such as "should" and "must." Female speech may contain more uncertainty and qualification, which some consider less dominant and powerful (Lakoff 1973). This suggests that one can display dominance and power through avoidance of uncertainty and qualification (Blankenship and Holtgraves 2005; Zhou et al. 2004b). Dominance is also correlated with being animated and open, implying that dominance is associated with greater exhibition of positive or negative emotions. Dominant individuals tend to make their emotions transparent to others (Burgoon et al. 1998). Females have higher emotional states when recalling past experiences, while in decision-making situations, women who are less dominant, tend to use more neutral language, rather than affective language (Zhou et al. 2004b). However, many of these findings relate to dyadic communications, where turn taking is assumed.

In groups, unlike dyadic communication, group members are held less accountable to keep the conversations going because the responsibility to contribute is shared among more individuals. Also, when subgroups exist in a group, subgroup members may attempt to control what information is communicated by other subgroup members and how others behave, along with self-impression management (Vernham et al. 2016). Therefore, whether the same linguistic features predict perceived dominance in a group conversational setting remains to be explored and is the subject of this study.

Deception and Language

Deception naturally involves the use of language to misdirect, conceal, and persuade. Behavioral theory suggests that the verbal content of deceivers differs from that of truth-tellers due to negative feelings associated with deception, increased cognitive effort, and attempts to control the narrative (Vrij 2008). Although deceivers may employ dominant strategies to persuade their audience (Zhou et al. 2004b), many verbal cues of deception are inversely related to those of dominance—especially those related to cognitive load. Dominant language is associated with assertiveness, authority, and self-confidence; however, lies tend to increase cognitive load, which manifests in fewer words and less lexical diversity (Hauch et al. 2015). When considering the emotional and cognitive aspects of deception, the language used by deceivers appears to be submissive. For instance, a deceiver may speak generally in an effort to provide fewer details, but equivocation is generally viewed as non-dominant (Buller et al. 1994).

Deceivers may limit their speaking time (DePaulo et al. 2003) in an effort to reduce attention or minimize verbal content that may provide additional signals of their deceit, which in turn may reduce perceptions of dominance. This effect may be moderated by factors such as motivation to succeed or time to prepare (Sporer and Schwandt 2006). Sporer and Schwandt (2006) found that deceivers who had more time to prepare a response talked for a shorter duration compared to truth-tellers, but this did not occur with short preparation time. Since deceivers in groups are not forced to engage the same way as deceivers in dyadic interaction, they have more time to prepare a response, thus, the response duration should be shorter.

Further, individuals who speak with fewer disfluencies, hedging terms and permission-seeking have their messages viewed as more persuasive and powerful (Blankenship and Holtgraves 2005). When deception requires greater cognitive resources than telling the truth, liars tend to show increased levels of speech disfluencies in the form of filled pauses (e.g. "uh", "um", "ah"; Vrij et al. 2008) and repeated words and phrases (DePaulo et al. 2003). Deceivers also tend to use more hedging and uncertain language (e.g. "may", "could", etc.) in an effort to introduce distance or vagueness into their statements (Burgoon et al. 2016; Zhou et al. 2004a).

Sentiment is also an important component of language that is related to both dominance and deception. As stated previously, dominant individuals tend to be more emotionally transparent in terms of both positive and negative emotions, which results in more affective terms in dominant language than submissive language. Deceivers also use emotional language, but the sentiment is generally negative (DePaulo et al. 2003; Hauch et al. 2015). This may be related to the negative emotions of guilt or fear felt when lying (Vrij 2008). Next, we will discuss the vocalic elements of dominance and deception and their relevance in group settings.

Vocalics

Vocalics or paralinguistics refer to characteristics of the voice such as pitch, loudness, and speaking rate (Burgoon et al. 2009). Research generally supports the idea that voice behaviors are closely tied to physiological processes of the human body (Juslin and Scherer 2005). The voice is increasingly used to automatically identify phenomena such as affect (e.g. Dhall et al. 2015; Jin et al. 2015), personality (e.g. Alam and Riccardi 2014; Vinciarelli and Mohammadi 2014), medical conditions (e.g. Bayestehtashk et al. 2015; Bone et al. 2014), and deception (e.g. Elkins et al. 2012; Rockwell et al. 1997; Twyman et al. 2015).

Air movement from the lungs through the vocal folds, nasal, and oral cavities is responsible for sound production during speech (Juslin and Scherer 2005). Factors such as the vocal fold length and tension are responsible for the characteristics of the sounds produced by the vocal folds, nasal folds and oral cavities. For instance, males have longer vocal folds compared to females, leading to a lower vocal pitch. Psychological factors such as arousal also influence sound production. High arousal is associated with more muscle tension, which leads to increased vibration in vocal folds causing a higher vocal pitch (Juslin and Scherer 2005).

The fundamental frequency (F_0) of the voice is interpreted as pitch. A higher pitch is variously interpreted as having more positive emotions (Oster and Risberg 1986) or as being associated with more arousal (Murray and Arnott 1993). Males with a lower pitch (Puts et al. 2007) and females with a higher pitch (Borkowska and Pawlowski 2011) are rated as more attractive. However, with respect to females, pitch has a ceiling effect where too high a pitch is rated as unattractive, possibly due to an association with adolescence (Borkowska and Pawlowski 2011).

Pitch variability is also a commonly studied characteristic of the voice. When people use greater pitch variability during speech, they are perceived as more benevolent (Brown et al. 1973), persuasive (Mehrabian and Williams 1969), and credible (Burgoon et al. 1990). Voice loudness, which reflects the amplitude of the audio signal (Eyben 2013), is also associated with persuasiveness (Mehrabian and Williams 1969).

Dominance and Voice

Dominance is the dynamic process of exerting power through relational and behavioral acts (Dunbar 2004). Behaviors perceived as more dominant include less positive affect, more eye gaze, more gesturing, more bodily openness, and more interruption during speech (Hall et al. 2005). Cheng et al. (2016) suggest that the voice is a flexible system used to establish dominance or submissiveness depending on ecological conditions. That is, unlike physical size, the voice is an alterable cue that can be adjusted depending on the need to be submissive or dominant in a given situation. They found that lower vocal pitch at the onset of an interaction positively influences the social ranking of an individual. Compared to a higher pitch, lower pitch is viewed as assertive, power seeking, and related to leadership and control (Cheng et al. 2016). Men and women with lower vocal pitch are perceived to be more dominant (Hall et al. 2005). Males who perceive themselves to be more (less) physically dominant to a male counterpart tend to lower (raise) their pitch during interactions (Puts et al. 2006). Although pitch is perhaps the most studied vocal characteristic of dominance, vocal loudness, vocal variability and speech rate are also found to correlate with perceptions of dominance (Hall et al. 2005). Greater loudness, more variability, and a faster speech rate are viewed as more dominant.

A perhaps lesser-studied vocal correlate of dominance is voice hoarseness. Voice hoarseness is quantified in terms of harmonic-to-noise ratio (HNR). HNR can be interpreted as a measure of voice quality. When asked to portray dominant voices, women have a tendency to lower their HNR, producing a more hoarse voice (Hughes et al. 2014). Hughes et al. also found that lower jitter and shimmer are rated as more dominant in females when judged by other females, and men rated women with louder voices as more dominant. The same trends were not found when men were judged as dominant.

Speaking time, a measure of activity (Pentland 2004) is another important behavioral feature of the voice. Speaking time in a group is highly correlated with ratings of dominance and this relationship is linearly related to group size – i.e. group size is positively correlated with dominance and speaking time (Mast 2002). Speaking time is commonly used in machine learning models to predict dominance (e.g. Hung et al. 2008; Jayagopi et al. 2009; Sanchez-Cortes et al. 2013). Sanchez-Cortes et al. (2013) argue that verbal cues, including speaking time, are better predictors of dominance than visual cues.

Deception and Voice

Deception is viewed as an arousing (Ekman and Friesen 1969) and cognitively complex process (Zuckerman et al. 1981). The strong connection between internal human processes and voice characteristics makes paralinguistic features useful signals for deception detection. Both the excitement of lying and the cognitive load that occurs in the production of deceit affect the characteristics of the voice. The leakage hypothesis (Ekman and Friesen 1969) posits that deception is associated with physiological emotions such as fear, nervousness, and even exhilaration (i.e., duping delight). Depending on the circumstances surrounding deceit, emotions can range from negative to positive; the high arousal of deception causes vocal folds to tighten which then causes variations in vocal pitch, vocal tension, and vocal quality (DePaulo et al. 2003).

Reporting what actually occurred as a narrative (truth-telling) is generally an easier task than manufacturing events, making deception a more cognitively complex task than truth-telling (Vrij et al. 2006; Zuckerman et al. 1981). The cognitive factors involved in the production of deception influences voice characteristics in the form of disturbances, delayed responses, and shorter message duration (Vrij 2008). This is especially true in situations where a deceiver is *on-the-spot* and must manufacturer a lie in real-time.

Other paralinguistic indicators of deception include a slower tempo (Mehrabian 1971), increased (Buller and Aune 1987) or decreased (Mehrabian 1971) loudness, and lower voice quality (e.g. lower harmonic-to-noise ratio) (Elkins et al. 2012). Like all cues to deception, the presence and discriminant ability of vocal cues vary based on extraneous factors (Vrij 2008). A liar's motivation to succeed, for instance, changes the cues produced (Burgoon 2005). Deception in an adversarial group setting is understudied, with different moderating factors that can possibly influence cues to deception.

Methodology

To investigate the linguistic and vocalic cues associated with dominance and deception in adversarial group settings, we followed the same methodology outlined in other chapters of this book: subjects participated in an adaptation of the popular Mafia game, with some players randomly assigned the role of villager (truthful) and others randomly assigned the role of spy (deceiver). During game play, participants (5 to 8 participants per game) were seated at desks arranged in a circular fashion and equipped with a laptop computer. Each laptop had a front-facing camera, which recorded the audio and video of the participant seated at the desk. The current study is concerned with the audio captured from the computers. We extracted linguistic and vocalic features from the recorded audio following a multi-step process that combined manual and automated methodologies. The following sections contain a description of the features we selected, and the specific techniques we used to extract the features.

Linguistic and Vocalic Features Selection

Although a wide variety of language and voice features can be generated, we selected a subset based on those identified in our literature review as being relevant to dominance and deception. Some features associated with dominance and deception were omitted from the current analysis due to environmental factors that prohibited feature extraction. For instance, response latency is commonly associated with deception, but since the experiment involved turn taking in a group format, the conditions restrict an accurate calculation of the measure. Tables 6.1 and 6.2 below outline the language and voice measures selected for analysis.

Linguistic and Vocalic Feature Extraction

The feature extraction process began by converting audio recordings to text transcriptions. For this, we used IBM's Watson Speech-to-Text service (IBM 2018). Considering the proximity of each recording device, a single laptop recording captured the speech from all study participants during a single game. We used these duplicate recordings to our advantage. All recordings from each game were processed with the automated speech recognition (ASR) tool. This produced multiple

Table 6.1 Language Measures

Measure Name	Definition
# of Words	Total words spoken by a participant for a given time window
# of Turns-at-Talk	Number of times a participant spoke for a given time window
Dominance Ratio	Ratio of dominant turns-at-talk to total number of turns-at-talk
Disfluency Ratio	Ratio of repeat phrases and filled pauses to the total number of words. Filled pauses are transcribed as "%HESITATION" by IBM Watson Speech-to-Text
Polarity score	The compound polarity score as computed by the VADER sentiment algorithm in NLTK (Hutto and Gilbert 2014).
Hedging Ratio	Ratio of number of hedging and uncertainty terms to total number of words

Table 6.2 Voice Measures

Measure Name	Definition
F_0 (pitch) Mean	The lowest frequency of a periodic waveform.
F_0 (pitch) Std	
Loudness-Mean	Subjective perception of sound pressure.
Loudness-Std	
HNR-Mean	The harmonic-to-noise ratio (HNR) is the proportion of harmonic sound to
HNR-Std	noise in the voice measured in decibels.
Jitter-Mean	Jitter is a measure of period-to-period fluctuations in fundamental frequency.
Jitter-Std	
Shimmer-Mean	Shimmer measures the variability of the amplitude value.
Shimmer-Std	
Turn-at-talk Duration	Duration in seconds of a turn-at-talk

transcriptions for each game. We then used Recognizer Output Voting Error Reduction (ROVER; Fiscus 1997) to merge multiple transcripts into a single transcript. ROVER uses a voting strategy across all same-game transcription to produce a final output and is found to reduce the word error rate of transcription. Along with a text transcription of each game, this process also produced word-level timestamps.

Although the ASR tool drastically decreases the transcription time compared to human-transcription, it is unable to distinguish between speakers. Output from the previous step contained all spoken words without the speaker identified. To identify speakers, we used a manual process where research assistants watched each video recording and coded the speaker in the transcription. The output of this effort produced a final transcription, which included the speaker and timestamp of each spoken word.

We extracted linguistic features from Table 6.1 using SPLICE, a linguistic cue extraction tool that uses word counts, part-of-speech tags, and dictionaries to return language measurements (Moffitt and Giboney 2011). Dominant turns-at-talk are those which contain phrases like "you must" or "I can", and the ratio is computed with the number of dominant turns-at-talk divided by the total number of turns-at-talk for a player in a time interval. We use the VADER compound polarity measure to calculate sentiment (Hutto and Gilbert 2014). The VADER algorithm is sophisticated enough to identify negations (e.g. "not good" reflects negative sentiment). The result is a score between −1 and +1 for each turn-at-talk, with +1 (−1) representing a strong positive (negative) sentiment, and a score of 0 for a statement with neutral sentiment. Since dominant individuals may use more extreme emotional language (strongly positive or strongly negative), we use the absolute value of the VADER score. Therefore, scores closer to zero indicate neutral language, and scores closer to one indicate affective language. We also extract the ratio of words that indicate hedging, and disfluency ratio. The disfluency ratio includes repeat phrases (e.g. "I think that… that is a good idea.") and filled pauses (e.g. "um", "uh", etc.) and is given by:

$$\frac{\text{Repeat Phrases} + \text{Filled Pauses}}{\text{Total Number of Words}}.$$

To extract vocalic measures, audio for each participant was segmented into turn-at-talk (TaT) clips based on transcription timestamps. Each audio clip was then processed with OpenSmile (Eyben & Schuller, 2014) to extract vocalic data. In some cases, audio files contained noise such as laughter or multiple voices at one time, which produced abnormal voice measures. An outlier ranking probability measure was derived using the extracted paralinguistic features to identify noisy audio segments. TaT segments with an outlier rank above the 85th quantile were removed from the dataset. TaT segments less than 1 second were also excluded from the dataset.

Vocalic and linguistic measures were averaged across three timeframes: T1) introduction, T2) Rounds 1 and 2, and T3) all remaining rounds. This method of aggregation allowed us to compensate for differing game lengths. It also created a baseline period (T1) where game role had not yet been assigned. TaT aggregation by timeframe produced three measures for each feature for each participant. The number of rounds per game varied from 2 to 8. To ensure the analysis included three distinct timeframes, games/players with only 2 rounds were removed.

Participants

For this analysis, we used a subset of culturally homogeneous games, meaning only games where all participants identified as being from the same country were analyzed. In total, 22 games involving 162 (Male = 68; Female = 94) participants were analyzed. Homogeneous games were selected from the Southwestern US (2 games; n = 14), Western US (2 games; n = 15), Northeastern US (1 game; n = 7), Israel (3 games; n = 20), Singapore (5 games; n = 37), Hong Kong (3 games; n = 23), Fiji (3 games; n = 22), and Zambia (3 games; n = 24). The number of participants per game ranged from 6 to 8.

Villagers won 13 out of the 22 homogeneous games. The villager's winning rate drops to nearly half (47/95) when all games are considered. Among the 162 players, 63 were assigned to be spies and 99 were villagers. The average age of the players in the homogeneous games is 22.13 years old, with a standard deviation of 3.71 years old (two participants did not report their age). 86 players (54.1%) had played a similar Mafia game before the experiment, and 78 (48.4%) were native English speakers.

For players who reported their ethnicities, 46 (49.5%) were Asian, 30 (32.3%) were white, 5 were Latin/Hispanic and 2 were Israeli, and 9 players identified themselves as multiracial. There were 69 players that did not report their ethnicities.

Analysis & Results

In this section, we report the modeling results of linguistic and vocalic features on dominance and deception respectively.

Dominance Analysis

Two mixed-effects models were specified to analyze the relationship between dominance and vocalic features and between dominance and linguistic features. Model 1 included vocalic features and Model 2 included linguistic features. For both models, dominance rating was the dependent variable. Control variables for each model included: timeframes (T1 = Introduction, T2 = Rounds 1 and 2, and T3 = All rounds after round 2), game role (Spy/Villager), gender (Male/Female), previous game experience (Yes/No), English native speaker (Yes/No), and game status (an indicator of which side (villager or spies) is leading the game during each timeframe). A game identifier was set as the random effect.

For the vocalic models, average level and standard deviation of five vocalic features were set as independent variable. Features include fundamental frequency (F_0), loudness, harmonics-to-noise ratio (HNR), jitter and shimmer of the player's acoustic signals. We also include an indicator of the average length of utterance (e.g. turns-at-talk duration) as an independent variable.

For the linguistic model on dominance, cues include dominance ratio, number of words, number of sentences, the absolute value of the sentiment score, the ratio of words that indicate hedging and the disfluency ratio. More specifically, the variable of disfluency is calculated based on the original disfluency ratio in the SPLICE output but also has hesitations. We use the absolute value of sentiment score of an utterance instead of the original numerical value since literature suggests that extremity of sentiment may be indicative to dominance. The disfluency ratio includes repeat phrases (e.g. "I think that… that is a good idea.") and filled pauses (e.g. "um", "uh", etc.).

Peer ratings of dominance measures were collected periodically throughout game play. Each player evaluated all other players level of dominance before the game and after rounds 2, 4, 6, and 8. The measure used for analysis is the average dominance rating provided by participants assigned to the 'villager' role. Measures provided by participants assigned to the role of 'spy' were excluded from the calculation because spies were informed of the roles of other players before the beginning of round 1.

Table 6.3   Baseline Model V.S. Full Vocalic Model

	Dependent variable: Dominance Score		
	Baseline Model	Simplified Model	Full Model
	β(SE)	β(SE)	β(SE)
T3	0.266(−0.12)**	0.207(−0.124)*	0.192(−0.124)
T2	0.081(−0.124)	0.033(−0.123)	0.02(−0.122)
Game Role	0.165(−0.132)	0.107(−0.126)	0.097(−0.126)
Gender	0.220(−0.084)***	0.058(−0.142)	0.031(−0.143)
Game Experience	0.157(−0.101)	0.155(−0.097)	0.158(−0.098)
Native English Speaker	0.196(−0.109)*	0.16(−0.106)	0.158(−0.107)
Game Status	−0.018(−0.038)	−0.024(−0.037)	−0.025(−0.037)
TaT duration		0.020(−0.007)***	0.019(−0.008)**
F_0-mean		−0.004(−0.002)*	−0.004(−0.002)
F_0-Sd		0.007(−0.004)*	0.006(−0.004)
Loudness-mean		0.221(−1.01)	0.026(−1.022)
Loudness-Sd		2.663(−1.323)**	2.412(−1.378)*
HNR-mean		0.004(−0.002)**	0.004(−0.002)**
HNR-Sd		−0.011(−0.005)**	−0.010(−0.005)**
Jitter-mean			0.727(−4.357)
Jitter-Sd			1.99(−2.852)
Shimmer-mean			8.547*(−5.027)
Shimmer-Sd			−2.981(−4.171)
Game Role * T3	−0.733(−0.188)***	−0.658(−0.18)***	−0.675(−0.178)***
Game Role * T2	−0.244(−0.193)	−0.194(−0.184)	−0.193(−0.183)
Constant	3.186(−0.148)***	2.872(−0.423)***	2.154(−0.57)***
Observations	388	388	388
Log Likelihood	−462.364	−440.736	−438.058
AIC	948.728	919.473	922.117
BIC	996.26	994.732	1013.22

Note: *p < 0.1; **p < 0.05; ***p < 0.01

Vocalic-Dominance Results

We compare a baseline model, which does not contain any vocalic features, with the full vocalic model, which includes all 11 vocalic features, and a simplified model which omits jitter and shimmer measures. The results are in Table 6.3.

In the model containing only the control variables, the third timeframe (T3) ($\beta = 0.266$, $p < 0.05$) and gender ($\beta = 0.22$, $p < 0.01$) had significant effects on the perceived dominance as separate independent variables in the baseline model. Further, the interaction term of game role and timeframe (T3) was also significant ($\beta = -0.733$, $p < .01$) in the baseline model, which means the perceived dominance in the game after round 2 is significantly lower for spies than villagers.

The full model included all measures related to dominance. The standard deviation of loudness, mean ($\beta = 2.412$, $p < 0.1$) and standard deviation ($\beta = -0.010$, $p < 0.05$) of HNR, mean of shimmer ($\beta = 8.547$, $p < 0.1$), and turns-at-talk duration ($\beta = 0.019$, $p < 0.05$) had significant relationships with a player's perceived dominance. In the full vocalic model timeframe and gender became insignificant. The interaction term of time frame (T3) and game role was still significant ($\beta = -0.658$, $p < .01$) and consistent with the baseline model.

Table 6.3 also includes a simplified model, which excludes jitter and shimmer. Jitter, shimmer and fundamental frequency are closely related acoustic measures. Jitter is a measure of period-to-period fluctuations in fundamental frequency while shimmer measures the variability of the amplitude value. It's reasonable to assume a certain degree of co-linearity between the three variables. To mitigate the issue of multi-collinearity, we drop the four vocalic features associated with jitter and shimmer and build a simplified vocalic model.

Among the control variables, T3 has a significant ($\beta = 0.207$, $p < 0.1$) effect on the player's perceived dominance. Further, the interaction term of T3 and game role is also significant ($\beta = -0.658$, $p < 0.01$), indicating that as the Mafia game evolves, villagers' perceptions of spies' dominance decreases. However, the game role alone does not significantly influence a player's perceived dominance.

In the simplified model, both the average level ($\beta = -0.004$, $p < 0.1$) and standard deviation ($\beta = 0.007$, $p < 0.1$) of the fundamental frequency of the player are significant. However, only the variation (standard deviation) of the loudness is significant ($\beta = 2.663$, $p < 0.05$). HNR is significant both for its mean ($\beta = 0.004$, $p < 0.05$) and standard deviation ($\beta = -0.011$, $p < 0.5$). Further, the turn-at-talk duration is also significant ($\beta = 0.020$, $p < 0.01$). Compared to the baseline model and the full vocalic model, the simplified vocalic model has better overall, which is indicated by a higher log likelihood and smaller AIC and BIC.

Linguistic Dominance Results

Similar to the vocalic models, we compared a baseline model with a linguistic model, which considered six linguistic measures. The results are in Table 6.4.

The significances and directions of the beta coefficients in the baseline linguistic model are consistent with those of the baseline vocalic model, since they were fitted with overlapping samples. The only exception is the control variable English Native Speaker, which is marginally significant ($\beta = 0.196$, $p < 0.1$) in the baseline vocalic model but insignificant in the baseline linguistic model.

Among all linguistic cues included in the model, only dominance ratio ($\beta = 0.733$, $p < 0.1$) and the number of words ($\beta = 0.291$, $p < 0.01$) had significant effects on the player's perceived dominance. Both a higher level of dominance ratio and a larger number of words indicate a higher level of perceived dominance.

Among the control variables, T3 ($\beta = -0.204$, $p < 0.1$) had a significant effect on the perceived dominance, and furthermore, the interaction term of T3 and Game

Table 6.4 Baseline Model V.S. Linguistic Model

	Dependent variable: Dominance Score	
	Baseline Model	Linguistic Model
	$\beta(SE)$	$\beta(SE)$
T3	0.240*(0.125)	−0.204*(0.123)
T2	−0.019 (0.131)	−0.167 (0.118)
Game Role	0.221 (0.138)	0.173 (0.120)
Gender	0.322*** (0.085)	0.132* (0.077)
Game Experience	0.130 (0.101)	0.082 (0.090)
Native English Speaker	0.132 (0.113)	0.014 (0.105)
Game Status	−0.008 (0.037)	−0.008 (0.033)
Dominance Ratio		0.733* (0.395)
Number of Words		0.291*** (0.105)
Number of Sentences		0.160 (0.111)
|Polarity|		−0.243 (0.412)
Hedge Ratio		−0.269 (1.093)
Disfluency Ratio		−0.815 (0.916)
T3 * Game Role	−0.789*** (0.197)	−0.687*** (0.171)
T2* Game Role	−0.307 (0.198)	−0.235 (0.173)
Constant	3.128*** (0.148)	3.511*** (0.173)
Observations	409	409
Log Likelihood	−508.373	−455.860
AIC	1040.746	947.720
BIC	1088.910	1019.967

Note: *p < 0.1; **p < 0.05; ***p < 0.01

Role ($\beta = -0.687$, $p < 0.01$) is also significant. Though the presence of significances is consistent with the baseline model, the direction of the beta coefficient for T3 ($\beta = -0.204$, $p < 0.1$) as an independent variable is opposite to the T3 coefficient ($\beta = 0.240$, $p < 0.1$) in the baseline model. Gender ($\beta = 0.322$, $p < 0.01$) also had a significant effect on the perceived dominance.

The linguistic model also outperforms its baseline counterpart with respect to data fitness, which is indicated by a higher log likelihood, smaller AIC and BIC. Next, we analyzed the effect of deception on the vocalic and linguistic measures.

Deception Analysis

To analyze the relationship between behavioral characteristics and deception, mixed-models were again specified using the same control variables, interactions, and random effects outlined above. However, each vocalic and linguistic measure

Table 6.5 Vocalic Deception Results

	Dependent Variable β (SE)						
	FF-mean	FF-Std	Loudness-mean	Loudness-Std	HNR-mean	HNR-Std	TaT Duration
T2	5.472 (3.430)	5.724*** (1.579)	0.034*** (0.013)	0.031*** (0.010)	1.947 (3.997)	0.731 (1.296)	−3.047*** (0.829)
T3	9.124*** (3.310)	8.492*** (1.525)	0.050*** (0.013)	0.045*** (0.010)	−0.022 (3.865)	1.408 (1.253)	−3.943*** (0.801)
Role	**−0.850 (3.647)**	**−1.248 (1.679)**	**0.005 (0.014)**	**−0.003 (0.011)**	**6.981 (4.250)**	**−0.551 (1.378)**	**1.727* (0.881)**
Gender	−57.016*** (2.286)	−16.802*** (1.064)	0.005 (0.009)	0.016** (0.007)	−8.262*** (2.720)	−1.570* (0.882)	1.014* (0.559)
Experience	−3.796 (2.613)	−1.784 (1.277)	0.0002 (0.011)	0.002 (0.009)	−6.654* (3.408)	−4.459*** (1.109)	−1.327** (0.672)
English	8.072*** (2.756)	0.651 (1.384)	0.024* (0.012)	0.025*** (0.009)	−2.779 (3.809)	0.004 (1.243)	0.522 (0.729)
Status	−0.556 (1.004)	−0.451 (0.486)	0.0003 (0.004)	0.001 (0.003)	−0.547 (1.277)	0.251 (0.415)	0.536** (0.255)
T2*Role	**−2.873 (5.352)**	**1.052 (2.462)**	**0.002 (0.020)**	**−0.005 (0.016)**	**−0.381 (6.228)**	**0.176 (2.019)**	**−2.624** (1.292)**
T3*Role	**−2.166 (5.211)**	**−0.976 (2.397)**	**0.002 (0.020)**	**−0.006 (0.016)**	**−0.189 (6.066)**	**0.021 (1.966)**	**−3.055** (1.258)**
Intercept	166.803*** (3.385)	62.225*** (1.852)	0.407*** (0.028)	0.180*** (0.014)	4.821 (10.044)	23.678*** (4.705)	11.018*** (0.983)
Observations	388	388	388	388	388	388	388
Log-Likelihood	−1741	−1448	391	495	−1832	−1404	−1199
AIC	3507	2921	−759	−966	3688	2832	2422
BIC	3555	2969	−712	−919	3736	2879	2470

Note: *p < 0.1; **p < 0.05; ***p < 0.01

was specified as the dependent variable in separate models and game role (villager or spy) was set as an independent measure. Results are in Tables 6.5, 6.6 and 6.7.

Among the 11 vocalic features included in the full vocalic model, only the turns-at-talk duration is significantly ($\beta = 1.727$, $p < 0.1$) influenced by the role of the player. That is to say, keeping the levels of all the other control variables the same, spies appeared to speak longer than villagers. Further, for turns-at-talk duration, the interactions of time frame and game role were also significant, which means that the spies and villagers not only behaved differently on average cross the whole game, but spies reduce their turn-at-talk duration as the game proceeds.

Table 6.6 Vocalic Deception Results Continued

	Dependent Variable β (SE)			
	Jitter-mean	Jitter-Sd	Shimmer-mean	Shimmer-Std
T2	0.003*	0.005*	0.003	0.002
	(0.002)	(0.003)	(0.002)	(0.002)
T3	0.004**	0.004	0.004**	0.001
	(0.002)	(0.003)	(0.002)	(0.002)
Role	**−0.001**	**−0.001**	**0.003**	**0.004**
	(0.002)	**(0.003)**	**(0.002)**	**(0.002)**
Gender	0.003**	0.003*	0.011***	0.021***
	(0.001)	(0.002)	(0.001)	(0.002)
Experience	−0.001	−0.003	0.001	−0.001
	(0.001)	(0.002)	(0.002)	(0.002)
English	0.001	0.001	0.003	0.002
	(0.002)	(0.002)	(0.002)	(0.002)
Status	−0.001	−0.001	0.0004	0.001
	(0.001)	(0.003)	(0.001)	(0.001)
T2*Role	**0.003**	**0.004**	**−0.002**	**−0.003**
	(0.003)	**(0.004)**	**(0.003)**	**(0.004)**
T3*Role	**0.004**	**0.004**	**−0.0004**	**−0.002**
	(0.003)	**(0.004)**	**(0.003)**	**(0.004)**
Intercept	0.029***	0.048***	0.109***	0.079***
	(0.002)	(0.003)	(0.002)	(0.002)
Observations	388	388	388	388
Log Likelihood	1166.503	1002.594	1168.696	1086.882
AIC	−2309.006	−1981.187	−2313.393	−2149.764
BIC	−2261.474	−1933.655	−2265.861	−2102.231

Note: $^*p < 0.1$; $^{**}p < 0.05$; $^{***}p < 0.01$

Linguistic Deception Results

We fitted the same mixed-effect models with the linguistic features being the dependent variable. However, Game Role did not significantly influence the dependent variable in any of the models. This includes the interaction terms between Time Frame and Game Role, which means, given the other control variables, there were no significant differences between spies and villagers with respect to the linguistic features considered in our models.

Table 6.7 Linguistic Deception Results

	Dependent Variable β (SE)					
	Dominance Ratio	Number of Words	Number of Sentences	Sentiment Score	Hedge Ratio	Disfluency
T2	−0.0002 (0.015)	0.283** (0.142)	0.350*** (0.133)	−0.048*** (0.014)	0.019*** (0.005)	−0.007 (0.006)
T3	0.004 (0.014)	0.825*** (0.135)	1.027*** (0.126)	−0.067*** (0.014)	0.014*** (0.005)	−0.020*** (0.006)
Role	**0.019 (0.016)**	**0.090 (0.151)**	**0.076 (0.141)**	**0.001 (0.015)**	**−0.002 (0.006)**	**−0.002 (0.007)**
Gender	−0.004 (0.010)	0.451*** (0.092)	0.432*** (0.086)	−0.008 (0.009)	−0.001 (0.003)	−0.004 (0.004)
Experience	0.011 (0.010)	0.134 (0.103)	0.136 (0.100)	0.008 (0.011)	0.003 (0.004)	0.002 (0.005)
English	0.011 (0.010)	0.290*** (0.111)	0.317*** (0.110)	−0.018 (0.013)	0.003 (0.004)	0.006 (0.005)
Status	−0.003 (0.004)	0.006 (0.038)	−0.022 (0.036)	−0.006 (0.004)	0.001 (0.001)	0.001 (0.002)
T2*Role	**−0.014 (0.023)**	**−0.163 (0.216)**	**−0.088 (0.202)**	**0.015 (0.022)**	**0.008 (0.008)**	**−0.001 (0.010)**
T3*Role	**0.006 (0.023)**	**−0.248 (0.212)**	**−0.292 (0.198)**	**−0.002 (0.022)**	**0.002 (0.008)**	**0.001 (0.010)**
Intercept	0.054*** (0.013)	−0.772*** (0.138)	−0.878*** (0.138)	0.178*** (0.017)	0.027*** (0.005)	0.049*** (0.011)
Observations	414	414	414	414	414	414
Log Likelihood	390.072	−542.132	−517.762	394.887	821.538	739.926
AIC	−756.145	1108.265	1059.524	−765.774	−1619.077	−1455.853
BIC	−707.834	1156.575	1107.834	−717.463	−1570.767	−1407.542

Note: $^{*}p < 0.1$; $^{**}p < 0.05$; $^{***}p < 0.01$

Discussion

Most research on linguistic and vocalic cues to dominance and deception focus on dyadic interactions, thus, what we understand about these cues is limited to the nuances of these types of interactions. The evaluation of group communication that has occurred is generally limited to triads, which does not capture the complexities of larger group interactions, especially with respect to adversarial environments. The goal of this study was to better understand linguistic and vocalic features associated with dominance in adversarial group settings and determine if an individual's role within a group affects their linguistic and vocalic behaviors.

To accomplish this, we developed and implemented a semi-automated process for language and voice feature extraction from adversarial group negotiation settings. Extracted features were analyzed with respect to dominance and game role (truthful or deceptive). Results indicate that dyads and groups share many of the same behavioral features related to dominance and deception, but specific differences exist.

Results of the vocalic model on dominance were largely consistent with the existing literature. In the full vocalic model, greater variability of loudness, higher voice quality (HNR), and longer turn-at-talk duration were associated with greater perceived dominance. In the simplified vocalic model, players with a lower pitch, higher voice quality, greater variability in pitch and loudness, and longer turn-at-talk duration were rated as more dominant. However, we did not find a significant correlation between loudness and perceived dominance as expected.

While the literature suggests dominance is associated with more extreme emotions and less hedging and hesitations, we did not find these relationships with regards to linguistic features. The number of words was a predictor of perceived dominance, consistent with previous research. Also, dominance ratio was positively correlated with perceived dominance as expected. Overall, vocalic features related to dominance in larger group settings are similar to those found in dyads. However, linguistic cues to dominance appear different in larger groups. Only two of the expected linguistic cues (number of words and dominance ratio) were significantly related to perceived dominance. This perhaps demonstrates that *it's not what you say, but how you say it*. Perceived dominance appears to be a function of overt characteristics of the voice oppose to semantic content.

Results of the vocalic models on deception showed that among the explored features, only longer turn-at-talk duration differentiated between game roles. The results indicated that spies (deceivers) tend to speak longer compared to villagers. Based, on prior literature, we expected measures related to pitch, loudness, and voice quality measures to differ between truth and deception. Similarly, none of the linguistic cues evaluated significantly differentiated between truth and deception.

The lack of findings related to deception suggests that cues to deception are different in larger groups. Although deception is thought to be more difficult in group settings, the inverse may potentially be true once a group has reached a larger size. Arguable, a large group setting offers advantages to deceivers compared to dyadic or smaller group settings. Groups are generally less emotionally charged compared to dyads (Moreland 2010) and this may be linearly related to group size. Since many of the cues to deception reside in the theory that high arousal is responsible for cue leakage (Ekman and Friesen 1969), cues to deceit may be muted or vary in the emotionally subdued environment of large groups.

Further, during large group communication, a deceiver may not be under the same level of scrutiny compared to dyadic or smaller group communication. At least in the case of this study, suspicion in a group setting is shared. This knowledge may reduce the arousal felt by deceivers since they do not feel like the primary target of suspicion, allowing the deceiver to act more naturally and reducing the occurrence of cues related to behavioral control (Burgoon 2005). A large group setting may also provide a deceiver more unobstructed time to consider strategy. Interactions between the many other group members provides time where a deceiver is not required to engage with the group and can observe behaviors and craft strategies. This also allows the deceiver to observe and accommodate/replicate the behaviors of other players, as a result, better obscuring their nefarious motives.

Conclusion

Group interactions are common in every-day life, but receive far less attention from researchers compared to dyadic interactions. The results of this study suggest that, in certain cases, results from dyadic interaction do generalize to group interactions; however, in other cases, they do not. This knowledge acts as both a warning and a call-to-action. We must not generalize verbal and nonverbal behaviors to group interactions without considering group factors that may alter behaviors.

Acknowledgement We are grateful to the Army Research Office for funding much of the work reported in this book under Grant W911NF-16-1-0342.

Funding Disclosure This research was sponsored by the Army Research Office and was accomplished under Grant Number W911NF-16-1-0342. The views and conclusions contained in this document are those of the authors and should not be interpreted as representing the official policies, either expressed or implied, of the Army Research Office or the U.S. Government. The U.S. Government is authorized to reproduce and distribute reprints for Government purposes notwithstanding any copyright notation herein.

References

Alam, F., & Riccardi, G. (2014). Fusion of acoustic, linguistic and psycholinguistic features for speaker personality traits recognition. In *2014 IEEE international conference on acoustics, speech and signal processing (ICASSP)* (pp. 955–959). IEEE. https://doi.org/10.1109/ICASSP.2014.6853738.

Baldwin, T. T., Bedell, M. D., & Johnson, J. L. (1997). The social fabric of a team-based M.B.A. program: Network effects on student satisfaction and performance. *Academy of Management Journal, 40,* 1369–1397. https://doi.org/10.5465/257037.

Báles, R. F., Strodtbeck, F. L., Mills, T. M., & Roseborough, M. E. (1951). Channels of communication in small groups. *American Sociological Review, 16,* 461–468. https://doi.org/10.2307/2088276.

Bayestehtashk, A., Asgari, M., Shafran, I., & McNames, J. (2015). Fully automated assessment of the severity of Parkinson's disease from speech. *Computer Speech & Language, 29,* 172–185. https://doi.org/10.1016/J.CSL.2013.12.001.

Blankenship, K. L., & Holtgraves, T. M. (2005). The role of different markers of linguistic powerlessness in persuasion. *Journal of Language and Social Psychology, 24,* 3–24.

Bone, D., Lee, C.-C., Black, M. P., Williams, M. E., Lee, S., Levitt, P., & Narayanan, S. (2014). The psychologist as an interlocutor in autism spectrum disorder assessment: Insights from a study of spontaneous prosody. *Journal of Speech Language and Hearing Research, 57,* 1162. https://doi.org/10.1044/2014_JSLHR-S-13-0062.

Borkowska, B., & Pawlowski, B. (2011). Female voice frequency in the context of dominance and attractiveness perception. *Animal Behaviour, 82,* 55–59. https://doi.org/10.1016/j.anbehav.2011.03.024.

Brown, B. L., Strong, W. J., & Rencher, A. C. (1973). Perceptions of personality from speech: Effects of manipulations of acoustical parameters. *The Journal of the Acoustical Society of America, 54,* 29–35. https://doi.org/10.1121/1.1913571.

Buller, D. B., & Aune, R. K. (1987). Nonverbal cues to deception among intimates, friends, and strangers. *Journal of Nonverbal Behavior, 11,* 269–290.

Buller, D. B., Burgoon, J. K., Buslig, A. L., & Roiger, J. F. (1994). Interpersonal deception VIII: Further analysis of nonverbal and verbal correlates of equivocation from the Bavelas et al. (1990) research. *Journal of Language and Social Psychology, 13,* 396–417.

Burgoon, J. K. (2005). The future of motivated deception and its detection. *Annals of the International Communication Association, 29,* 49–96.

Burgoon, J. K., Birk, T., & Pfau, M. (1990). Nonverbal behaviors, persuasion, and credibility. *Human Communication Research, 17,* 140–169. https://doi.org/10.1111/j.1468-2958.1990.tb00229.x.

Burgoon, J. K., Johnson, M. L., & Koch, P. T. (1998). The nature and measurement of interpersonal dominance. *Communications Monographs, 65,* 308–335.

Burgoon, J. K., Guerrero, L. K., & Floyd, K. (2009). *Nonverbal communication.* New York: Allyn & Bacon.

Burgoon, J. K., Mayew, W. J., Giboney, J. S., Elkins, A. C., Moffitt, K., Dorn, B., Byrd, M., & Spitzley, L. (2016). Which spoken language markers identify deception in high-stakes settings? Evidence from earnings conference calls. *Journal of Language and Social Psychology, 35,* 123–157.

Cartwright, D., & Zander, A. (1968). *Group dynamics* (3rd ed.). Oxford, UK: Harper + Row.

Cheng, J. T., Tracy, J. L., Ho, S., & Henrich, J. (2016). Listen, follow me: Dynamic vocal signals of dominance predict emergent social rank in humans. *Journal of Experimental Psychology: General, 145,* 536.

Cohen, A. L., Cash, D., & Muller, M. J. (2000). Designing to support adversarial collaboration. In *Proceedings of the 2000 ACM conference on computer supported cooperative work, CSCW '00* (pp. 31–39). New York: ACM. https://doi.org/10.1145/358916.358948.

DePaulo, B. M., Lindsay, J. J., Malone, B. E., Muhlenbruck, L., Charlton, K., & Cooper, H. (2003). Cues to deception. *Psychological Bulletin, 129,* 74.

DeSanctis, G., & Gallupe, R. B. (1987). A foundation for the study of group decision support systems. *Management Science.* https://doi.org/10.1287/mnsc.33.5.589.

Dhall, A., Ramana Murthy, O. V., Goecke, R., Joshi, J., & Gedeon, T. (2015). Video and image based emotion recognition challenges in the wild: Emotiw 2015. In *Proceedings of the 2015 ACM on international conference on multimodal interaction* (pp. 423–426). ACM.

Dunbar, N. E. (2004). Theory in progress: Dyadic power theory: Constructing a communication-based theory of relational power. *Journal of Family Communication, 4,* 235–248.

Ekman, P., & Friesen, W. V. (1969). Nonverbal leakage and cues to deception. *Psychiatry, 32,* 88–106.

Elkins, A. C., Derrick, D. C., & Gariup, M. (2012). The voice and eye gaze behavior of an imposter: Automated interviewing and detection for rapid screening at the border. In *Proceedings of the workshop on computational approaches to deception detection, EACL 2012* (pp. 49–54). Stroudsburg: Association for Computational Linguistics.

Eyben, F., Weninger, F., Gross, F., & Schuller, B. (2013). Recent developments in openSMILE, the Munich open-source multimedia feature extractor. *Proceedings of the 21st ACM International Conference on Multimedia,* 835–838. https://doi.org/10.1145/2502081.2502224.

Fiscus, J. G. (1997). A post-processing system to yield reduced word error rates: Recognizer output voting error reduction (ROVER). In *1997 IEEE workshop on automatic speech recognition and understanding proceedings. Presented at the 1997 IEEE workshop on automatic speech recognition and understanding proceedings* (pp. 347–354). https://doi.org/10.1109/ASRU.1997.659110.

Gallupe, R. B., Dennis, A. R., Cooper, W. H., Valacich, J. S., Bastianutti, L. M., Jay, F., & Nunamaker, J. (1992). Electronic brainstorming and group size. *Academy of Management Journal.* https://doi.org/10.5465/256377.

Giles, H., & Baker, S. C. (2008). Communication accommodation theory. In W. Donsbach (Ed.), *The international encyclopedia of communication.* Chichester: Wiley. https://doi.org/10.1002/9781405186407.wbiecc067.

Giordano, G. A., Stoner, J. S., Brouer, R. L., & George, J. F. (2007). The influences of deception and computer-mediation on dyadic negotiations. *Journal of Computer-Mediated Communication, 12,* 362–383. https://doi.org/10.1111/j.1083-6101.2007.00329.x.

Hall, J. A., Coats, E. J., & LeBeau, L. S. (2005). Nonverbal behavior and the vertical dimension of social relations: A meta-analysis. *Psychological Bulletin, 131,* 898.

Hare, A. (1981). Group size. *American Behavioral Scientist, 24,* 695–708.

Hauch, V., Blandon-Gitlin, I., Masip, J., & Sporer, S. L. (2015). Are computers effective lie detectors? A meta-analysis of linguistic cues to deception. *Personality and Social Psychology Review, 19,* 307–342.

Holtgraves, T. M., & Lasky, B. (1999). Linguistic power and persuasion. *Journal of Language and Social Psychology, 18,* 196–205.

Hosman, L. A. (1989). The evaluative consequences of hedges, hesitations, and intensifies: Powerful and powerless speech styles. *Human Communication Research, 15,* 383–406. https://doi.org/10.1111/j.1468-2958.1989.tb00190.x.

Hughes, S. M., Mogilski, J. K., & Harrison, M. A. (2014). The perception and parameters of intentional voice manipulation. *Journal of Nonverbal Behavior, 38,* 107–127. https://doi.org/10.1007/s10919-013-0163-z.

Hung, H., Jayagopi, D. B., Ba, S., Odobez, J.-M., & Gatica-Perez, D. (2008). Investigating automatic dominance estimation in groups from visual attention and speaking activity. In *Proceedings of the 10th international conference on multimodal interfaces, ICMI '08* (pp. 233–236). New York: ACM. https://doi.org/10.1145/1452392.1452441.

Hutto, C. J., & Gilbert, E. E. (2014). VADER: A parsimonious rule-based model for sentiment analysis of social media text. In *Eighth international AAAI conference on weblogs and social media. Presented at the eighth international conference on weblogs and social media.* Ann Arbor.

IBM. (2018). *Watson speech to text.* International Business Machines (IBM).

Itakura, H. (2001). *Conversational dominance and gender: A study of Japanese speakers in first and second language contexts.* Amsterdam/Philadelphia: John Benjamins Publishing.

Jayagopi, D. B., Hung, H., Yeo, C., & Gatica-Perez, D. (2009). Modeling dominance in group conversations using nonverbal activity cues. *IEEE Transactions on Audio, Speech, and Language Processing, 17,* 501–513. https://doi.org/10.1109/TASL.2008.2008238.

Jin, Q., Li, C., Chen, S., & Wu, H. (2015). Speech emotion recognition with acoustic and lexical features. In *2015 IEEE international conference on acoustics, speech and signal processing (ICASSP)* (pp. 4749–4753). IEEE. https://doi.org/10.1109/ICASSP.2015.7178872.

Juslin, P. N., & Scherer, K. R. (2005). Vocal expression of affect. In J. A. Harrigan, R. Rosenthal, & K. R. Scherer (Eds.), *New handbook of methods in nonverbal behavior research* (pp. 65–135). New York: Oxford University Press.

Kelleher, J. (2000). Review of traditional and collaborative models for negotiation. *Journal of Collective Negotiations in the Public Sector, 29,* 321–336.

Lakoff, R. (1973). Language and woman's place. *Language in Society, 2,* 45–79.

Linell, P., Gustavsson, L., & Juvonen, P. (1988). *Interactional dominance in dyadic communication: A presentation of initiative-response analysis.* Berlin/New York: Walter de Gruyter.

Marett, L. K., & George, J. F. (2004). Deception in the case of one sender and multiple receivers. *Group Decision and Negotiation, 13,* 29–44. https://doi.org/10.1023/B:GRUP.0000011943.73672.9b.

Mast, M. S. (2002). Dominance as expressed and inferred through speaking time. *Human Communication Research, 28,* 420–450. https://doi.org/10.1111/j.1468-2958.2002.tb00814.x.

McCullagh, P. D., & Landers, D. M. (1976). Size of audience and social facilitation. *Perceptual and Motor Skills, 42*, 1067–1070. https://doi.org/10.2466/pms.1976.42.3c.1067.

Mehrabian, A. (1971). Nonverbal betrayal of feeling. *Journal of Experimental Research in Personality, 5*, 64–73.

Mehrabian, A., & Williams, M. (1969). Nonverbal concomitants of perceived and intended persuasiveness. *Journal of Personality and Social Psychology, 13*, 37–58. https://doi.org/10.1037/h0027993.

Migdał, P. (2010). A mathematical model of the Mafia game. arXiv:1009.1031 [physics].

Moffitt, K.C., & Giboney, J. (2011). Structured Programming for Linguistic Cue Extraction (SPLICE).

Moreland, R. L. (2010). Are dyads really groups? *Small Group Research, 41*, 251–267.

Murray, I. R., & Arnott, J. L. (1993). Toward the simulation of emotion in synthetic speech: A review of the literature on human vocal emotion. *The Journal of the Acoustical Society of America, 93*, 1097–1108. https://doi.org/10.1121/1.405558.

Norton, R. (1983). *Communicator style: Theory, applications, and measure*. Beverly Hills: Sage.

Oade, A. (2011). What is an adversarial working relationship? In A. Oade (Ed.), *Working in adversarial relationships: Operating effectively in relationships characterized by little trust or support* (pp. 1–16). London: Palgrave Macmillan UK. https://doi.org/10.1057/9780230292390_1.

Oster, A., & Risberg, A. (1986). The identification of the mood of a speaker by hearing impaired listeners. *SLT-Quarterly Progress Status Report, 4*, 79–90.

Pentland, A. (2004). Social dynamics: Signals and behavior. In *Proceedings of the third international conference on developmental learning (ICDL'04)* (pp. 263–267). Salk Institute, San Diego. UCSD Institute for Neural Computation.

Pruitt, D. G. (1983). Strategic choice in negotiation. *The American Behavioral Scientist (pre-1986); Thousand Oaks, 27*, 167.

Puts, D. A., Gaulin, S. J., & Verdolini, K. (2006). Dominance and the evolution of sexual dimorphism in human voice pitch. *Evolution and Human Behavior, 27*, 283–296.

Puts, D. A., Hodges, C. R., Cárdenas, R. A., & Gaulin, S. J. C. (2007). Men's voices as dominance signals: Vocal fundamental and formant frequencies influence dominance attributions among men. *Evolution and Human Behavior, 28*, 340–344. https://doi.org/10.1016/j.evolhumbehav.2007.05.002.

Rockwell, P., Buller, D. B., & Burgoon, J. K. (1997). Measurement of deceptive voices: Comparing acoustic and perceptual data. *Applied PsychoLinguistics, 18*, 471–484.

Rude, S., Gortner, E.-M., & Pennebaker, J. (2004). Language use of depressed and depression-vulnerable college students. *Cognition & Emotion, 18*, 1121–1133. https://doi.org/10.1080/02699930441000030.

Sanchez-Cortes, D., Aran, O., Jayagopi, D. B., Schmid Mast, M., & Gatica-Perez, D. (2013). Emergent leaders through looking and speaking: From audio-visual data to multimodal recognition. *Journal on Multimodal User Interfaces, 7*, 39–53. https://doi.org/10.1007/s12193-012-0101-0.

Schweitzer, M. E., DeChurch, L. A., & Gibson, D. E. (2005). Conflict frames and the use of deception: Are competitive negotiators less ethical? *Journal of Applied Social Psychology, 35*, 2123–2149. https://doi.org/10.1111/j.1559-1816.2005.tb02212.x.

Sporer, S. L., & Schwandt, B. (2006). Paraverbal indicators of deception: A meta-analytic synthesis. *Applied Cognitive Psychology: The Official Journal of the Society for Applied Research in Memory and Cognition, 20*, 421–446.

Stogdill, R. M. (1974). *Handbook of leadership: A survey of theory and research, handbook of leadership: A survey of theory and research*. New York: Free Press.

Tausczik, Y. R., & Pennebaker, J. W. (2010). The psychological meaning of words: LIWC and computerized text analysis methods. *Journal of Language and Social Psychology, 29*, 24–54. https://doi.org/10.1177/0261927X09351676.

Thomas, E. J., & Fink, C. F. (1963). Effects of group size. *Psychological Bulletin, 60*, 371–384. https://doi.org/10.1037/h0047169.

Twyman, N. W., Proudfoot, J. G., Schuetzler, R. M., Elkins, A. C., & Derrick, D. C. (2015). Robustness of multiple indicators in automated screening systems for deception detection. *Journal of Management Information Systems, 32*, 215–245. https://doi.org/10.1080/0742122 2.2015.1138569.

Valacich, J. S., Dennis, A. R., & Nunamaker, J. F. (1992). Group size and anonymity effects on computer-mediated idea generation. *Small Group Research, 23*, 49–73.

Vernham, Z., Granhag, P.-A., & Giolla, E. M. (2016). Detecting deception within small groups: A literature review. *Frontiers in Psychology, 7*. https://doi.org/10.3389/fpsyg.2016.01012.

Vinciarelli, A., & Mohammadi, G. (2014). A survey of personality computing. *IEEE Transactions on Affective Computing, 5*, 273–291. https://doi.org/10.1109/TAFFC.2014.2330816.

Vrij, A. (2008). *Detecting lies and deceit: Pitfalls and opportunities*. Chichester: Wiley.

Vrij, A., Fisher, R., Mann, S., & Leal, S. (2006). Detecting deception by manipulating cognitive load. *Trends in Cognitive Sciences, 10*, 141–142.

Vrij, A., Mann, S. A., Fisher, R. P., Leal, S., Milne, R., & Bull, R. (2008). Increasing cognitive load to facilitate lie detection: The benefit of recalling an event in reverse order. *Law and Human Behavior, 32*, 253–265.

Weisfeld, G. E., & Linkey, H. E. (1985). Dominance displays as indicators of a social success motive. In S. L. Ellyson & J. F. Dovidio (Eds.), *Power, dominance, and nonverbal behavior* (pp. 109–128). New York: Springer Series in Social Psychology. Springer New York. https://doi.org/10.1007/978-1-4612-5106-4_6.

Williams, K. D. (2010). Dyads can be groups (and often are). *Small Group Research, 41*, 268–274. https://doi.org/10.1177/1046496409358619.

Yang, H.-L., & Tang, J.-H. (2003). Effects of social network on students' performance: A web-based forum study in Taiwan. *Journal of Asynchronous Learning Networks, 7*, 93–107.

Zajonc, R. B. (1965). Social Facilitation. *Science, 149*, 269–274.

Zhou, L., & Zhang, D. (2006). A comparison of deception behavior in dyad and triadic group decision making in synchronous computer-mediated communication. *Small Group Research, 37*, 140–164.

Zhou, L., Burgoon, J. K., Nunamaker, J. F., & Twitchell, D. (2004a). Automating linguistics-based cues for detecting deception in text-based asynchronous computer-mediated communications. *Group Decision and Negotiation, 13*, 81–106.

Zhou, L., Burgoon, J. K., Zhang, D., & Nunamaker, J. F. (2004b). Language dominance in interpersonal deception in computer-mediated communication. *Computers in Human Behavior, 20*, 381–402.

Zuckerman, M., DePaulo, B. M., & Rosenthal, R. (1981). Verbal and non-verbal communication of deception. In L. Berkowitz (Ed.), *Advances in experimental and social psychology* (pp. 1–59). New York: Academic.

Chapter 7
Attention-Based Facial Behavior Analytics in Social Communication

Lezi Wang, Chongyang Bai, Maksim Bolonkin, Judee K. Burgoon, Norah E. Dunbar, V. S. Subrahmanian, and Dimitris Metaxas

Introduction

Research shows that when humans communicate, more social meaning comes from nonverbal than verbal cues, and among the nonverbal modalities, the face is the one upon which people typically rely (Burgoon 2007). Facial expressions convey one's identity, display emotions, show status, give context, open or shut down conversation, signal approval, and reveal strength of conviction, among other things. People rely on facial cues to glean both intentional and unintentional meaning. With so much communicated by the face, it is natural that facial expressions have been investigated for possible cues to deception for decades (Burgoon 2007, 2017; DePaulo and Cooper 2003). With advances in computer vision has come the possibility of detecting facial movement variations on a more granular scale than the human eye can perceive, and with it, the discovery of deception indicators not normally directly detectable by human perception (Burgoon 2014; Tsiamyrtzis and Ekman 2007). Although much deception research has focused on the emotional potential of the face, searching for micro-level "leaked" indicators that betray concealed true emotions (Ekman 1975), the face can reveal far more signals related to

L. Wang · D. Metaxas (✉)
Rutgers University, New Brunswick, NJ, USA
e-mail: dnm@cs.rutgers.edu

C. Bai · M. Bolonkin · V. S. Subrahmanian
Dartmouth College, Hanover, NH, USA

J. K. Burgoon
University of Arizona, Tucson, AZ, USA

N. E. Dunbar
University of California, Santa Barbara, Santa Barbara, CA, USA

© Springer Nature Switzerland AG 2021   123
V. S. Subrahmanian et al. (eds.), *Detecting Trust and Deception in Group Interaction*, Terrorism, Security, and Computation,
https://doi.org/10.1007/978-3-030-54383-9_7

deception. It may reveal signs of cognitive effort and efforts to retrieve information from memory as a speaker attempts to formulate a believable verbal statement (Burgoon 2015). For example, blink patterns and lip presses may be associated with a speaker's thought processes. The face may signal not just internal emotional states such as fear or distress, but also affect directed toward another such as contempt or dislike. Nose flares and inauthentic smiles may signify these states. Communicators may also signal their attentiveness to others or their desire for a speaking turn (Burgoon 2005). Because people are aware that others' gaze is directed to the face, deceivers try to control their face and may, in the process, inadvertently over control it, producing a pattern of rigidity (Pentland 2017; Twyman 2014). Head movement, facial animation and gaze patterns may all reflect this "freezing" of activity. However, if deceivers have opportunities to rehearse, plan or mentally edit what they say, any temporary missteps may be repaired (Elkins 2015). Given the fluidity of facial expressions, temporal patterns can also be telling. For instance, blink patterns vary during versus after lying, and the onset and offset of smiles may differ by truth tellers versus liars.

In previous research, to automatically decipher the meaning in nonverbal human communication using computer vision methods, researchers first applied models inspired by communication theory. However, the underlying human defined features for the computer vision based analysis where incomplete due to non-linearity and the multi-scale nature of the problem. The recent use of neural nets, has addressed the discovery of the features associated with the computer vision-based analysis of nonverbal communication and has improved significantly the recognition of desired events during nonverbal communication such as truth telling.

In this paper, we develop a novel attention-based neural network (NN) approach in order to advance the state of the art in understanding inference in deep neural nets. Our novel approach discovers the frames in a video sequence and their content through AUs that contributed the most in the final inference of the neural net. This is done by employing a novel learning approach, at the various layers of the NN, to discover those pixels and related frames which are discriminative for the NN's class inference.

We train and test our novel approach on facial video collected from a version of the board game The Resistance. In this game players were randomly and secretly assigned to play deceivers (called "Spies"), or truth-tellers (called "Villagers"). The video-based facial data of the players were collected in various countries. Using our method the goal was to recognize who is a spy and who is a villager and also discover which frames and which facial expressions (AUs) contributed to the NN's class decision. Our approach demonstrates that with over 280 videos (2 hr. length each), we are on par with human recognition of spies vs. villagers. In addition, for the first time our NN can attend and discover the frames and associated facial action units (AUs) that contributed to the NN's class decision.

Related Work

Visualizing CNNs

A number of previous works have been proposed to visualize the internal representations offline in an attempt to better understand the model. In (Simonyan et al. 2013; Springenberg et al. 2014; Zeiler and Fergus 2014), they compute the gradient of the prediction w.r.t the specific CNN unit, i.e. the input image, to highlight the important pixels. Specifically, Simonyan et al. (2013) visualize partial derivatives of predicted class scores w.r.t. pixel intensities, while Guided Backpropagation (Springenberg et al. 2014) and Deconvolution (Zeiler and Fergus 2014) make modifications to raw' gradients that result in the better visualization. Despite producing fine-grained visualizations, these methods are not class-discriminative.

Erhan et al. (2009) synthesize the images to maximally activate a network unit and Mahendran and Vedaldi (2015), Dosovitskiy and Brox (2016) analyze the visual coding so as to invert latent representation. Although these can be high-resolution and class-discriminative, they visualize a model overall and not predictions for specific input images.

Our work is mainly inspired by recent works (Zhou et al. 2016; Selvaraju et al. 2017; Chattopadhay et al. 2018) addressing the class-discriminative attention maps. CAM (Zhou et al. 2016) generates the class activation maps highlighting the task relevant region by replacing fully-connected layers with convolution and global average pooling. A drawback of CAM is the low flexibility, which requires retraining of the classifiers and feature maps to directly precede softmax layers. Hence it is unable to be applicable to any feature layers. Grad-CAM (Selvaraju et al. 2017) is proposed to address this issue. Without retraining and changing network architecture, Grad-CAM generates the class activation maps by weighted combination of the feature maps in different channels. The weights are computed by the averaging of the gradient of the final prediction w.r.t the pixels in feature map. According to our observation, simple averaging is unable to measure the channel importance properly, which causes a large attention inconsistency among different feature layers. Grad-CAM++ (Chattopadhay et al. 2018) proposed a better class activation map by modifying the weight computation while its high computation cost of calculating the second and third derivatives makes it hard to be used to train the model.

Attention-Guided Network Training

There has been a number of recent works incorporating the model attention to guide CNN training in the vision researches. In (Zhang et al. 2018; Li et al. 2018; Wei et al. 2017; Chaudhry et al. 2017), they take advantage of the model attention to do weakly-supervised object localization and semantic segmentation. Specifically, with only image-level annotation, the attention of a well-trained classification model can highlight the important pixels in the original images. The model attention provides the object location information which reduces the burden of annotations in level of bounding-boxes and pixels.

Several works (Wang et al. 2017; Hu et al. 2017; Jetley et al. 2018; Woo et al. 2018) attempt to incorporate attention processing to improve the performance of CNNs in large-scale image classification. Wang et al. (2017) propose Residual Attention Network which modify the Residual Network (ResNet) (He et al. 2016) by adding hourglass net to the skip-connection to generate the attention masks to refine the feature maps. Hu et al. (2017) introduces a Squeeze-and Excitation module to exploit the inter-channel relationship, using the global average-pooled features to compute channel-wise attention. CBAM (Woo et al. 2018; Park et al. 2018) modifies Squeeze-and-Excitation module to exploit both spatial and channel-wise attention. More close to our work, Jetley et al. (2018) estimates the attention by taking feature maps at different stages in the CNN pipeline as input and outputting a 2D matrix of scores for each map. The output scores are used to predict the category. To our knowledge, we are the first to incorporate the class-specific attention to train the network for image classification, which is more closed to human perception than the attention mechanism generated in those previous work. And our proposed algorithm is a add-on module during training, without changing to the network architectures compared to those works.

Video Highlight Detection

Video highlight detection is highly related to our research topic since we intend to extract a brief synopsis containing segments of special interest from a video (Yao et al. 2016). Many earlier approaches have primarily been focused on highlighting sports videos. A latent SVM model is employed to detect highlights by learning from pairs of raw and edited videos (Sun et al. 2014). Success of deep learning also imparted improved performance in highlight detection (Yang et al. 2015). However, most of these techniques may not generalize well to web videos since they are either based on heuristic rules or require huge amount of human-crafted labelling data which are difficult to collect in many cases. In our The Resistance games, we only have video-level annotations of players' roles (Spy/Villager) without knowledge about when and where, in the untrimmed videos, players show the notable facial movements for the roles. Finding those movements are important for understanding human behaviors during communication. To achieve this goal, we incorporate the interpretation in the learning to discover those pixels and related frames which are discriminative and contributed the most to the NNs prediction for the players' roles.

Methodology

In this section, we describe the detail of how to extract the class-discriminative attention map for the videos. The procedure is illustrated by Fig. 7.1. Motivated by the work of Grad-CAM (Selvaraju et al. 2017) and Grad-CAM++ (Chattopadhay et al. 2018), we use the gradient to measure the importance of each feature map pixel to having the model classify the input image as class c. For the gradient of the class score Y c is computed by taking derivative w.r.t feature map Fk in k-th channels, i.e. $(\partial Y c)/\partial Fk)$. The pixel importance is denoted as $(\partial Y c)/(\partial Fijk)$. In (Selvaraju et al. 2017; Chattopadhay et al. 2018), the gradients are used to compute the channel-wise weights for combining the feature maps from different channels, generating the attention map of the last feature layer $A_{\text{Grad} - \text{CAM}}$:

$$A_{\text{Grad}-\text{CAM}} = ReLu\left(\sum \alpha_k^c F^k\right) \qquad (7.1)$$

, where α_k^c indicates the importance of the feature map F^k in the k-th channel. The weight α_k^c is a global average of pixel importance in the feature map:

$$\alpha_k^c = \frac{1}{Z}\sum_i\sum_j \frac{\partial Y^c}{\partial F_{ij}^k} \qquad (7.2)$$

, where Z indicates the total number of pixels in feature map F^k In (Chattopadhay et al. 2018), higher order derivatives (second and third) involved to compute the channel weights increase the computational costs.

Besides only generating the attention map of the last feature layer as in (Selvaraju et al. 2017; Chattopadhay et al. 2018), we compute the category-oriented attention map for the intermediate layers. In terms of the interpretability, we propose two

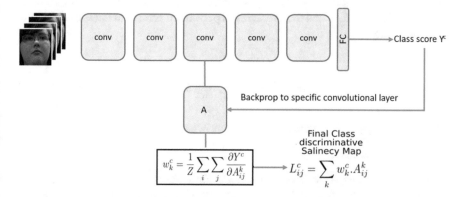

Fig. 7.1 The attention maps are generated via weighted combination of the feature maps at the specific layers. The weights measure the importance of the features, computed according to the gradients, where we take derivative of the class score w.r.t the feature maps

attention mechanisms for any feature layer with low computational cost, modeling the channel and pixel wise attention respectively. Then, we combine the model's channel and pixel wise attention to generate the final response map for the input video.

Channel-wise attention Different from Eq. 7.1 that the Grad-CAM uses the gradients of all the pixel to compute the channel weight, we only select the positive gradients and average them to obtain the channel-wise importance:

$$\alpha_k^c = \frac{1}{Z}\sum_i \sum_j ReLu\left(\frac{\partial Y^c}{\partial F_{ij}^k}\right) \tag{7.3}$$

The intuition is that the positive gradients model the pixels where the intensity increasing has positive impact on the final prediction score (Chattopadhay et al. 2018).

Substitute Eq. 7.3 to Eq. 7.1, we have the channel wise class-discriminative attention Ach.

$$\mathcal{A}_{ch} = \frac{1}{Z}ReLu\left(\sum_k\left(\sum_i\sum_j ReLu\left(\frac{\partial Y^c}{\partial F_{ij}^k}\right)\right)F^k\right) \tag{7.4}$$

Pixel-wise attention Attention proposed by the previous works (Zhou et al. 2016; Selvaraju et al. 2017; Chattopadhay et al. 2018) are computed in channel wise way, where the pixels within the same channel share the same weight for feature maps combination. Beside channel-wise attention, we also find that the pixel-wise attention demonstrates better guidance when training a model in low quality images. Specifically, each channel acts as an expert to vote the pixel importance in the attention map. In the feature map F^k, the pixel intensity is scaled by its importance measured as $\left(\partial Y^c\right)/\left(\partial F_{ij}^k\right)$ and the averaging is performed across channels to obtain the pixel-wise attention:

$$\mathcal{A}_{px} = ReLu\left(\frac{1}{K}\sum_k <\frac{\partial Y^c}{\partial F^k}, F^k>\right) \tag{7.5}$$

, where the $\left\langle\frac{\partial Y^c}{\partial F^k}, F^k\right\rangle$ indicates the element-wise multiplication between the gradient and feature maps.

The harmonic attention According to our observation, the pixel wise attention captures more high-frequency items and the channel-wise attention maps are smoother. Those two types of attention are complimentary where the A_{px} highlight the important pixels which are ignored by Ach due the low value averaged channel weights. Hence, we propose to combine the A_{px} and A_{ch}, generating the harmonic attention A. Empirically, applying the pixel-wise weighting first and then computing the channel-wise attention as Eq.7.3 achieves better performance. The proposed harmonic attention is formulated as:

$$A = \frac{1}{Z} ReLu\left(\sum_k \sum_i \sum_j ReLu\left(\frac{\partial Y^c}{\partial F_{ij}^k}\right) < \frac{\partial Y^c}{\partial F^k}, F^k > \right) \qquad (7.6)$$

Training a 3D Convolutional Neural Network for Spy Detection

We formulate the spy detection as a binary classification problem. Given a video sequence, we apply a 3D convolutional Neural Network (C3D) (Tran et al. 2015) to classify the player as spy or villager. Specifically, we crop the players' faces and the C3D takes a facial video clip as input, predicting the probability for the his/her role. The cropped face frames are normalized into the size of 112×112. In the C3D architecture, we design the model having 8 convolutions, 5 max-pooling, and 2 fully connected layers followed by a softmax layer. The 3D convolution kernels are $3 \times 3 \times 3$ with stride 1 in both spatial and temporal dimensions. The Number of filters are denoted in each box, as shown in Fig. 7.2. The 3D pooling layers are denoted from pool1 to pool5. All pooling kernels are $2 \times 2 \times 2$, except for pool1 is $1 \times 2 \times 2$. Each fully connected layer has 4096 output units. The output has 2 dimensions for binary classification. The model training and testing are conducted by using PyTorch and NVIDIA K80 XGPUs.

Experiments

Dataset

We tested our method on data from a real world game, where the goal is to examine deceivers' strategies and truth-tellers' deception detection abilities. Groups of participants were sought to play a board game adapted from Resistance, during which players in the roles of Villagers (truth-tellers) and Spies (deceivers) competed to win missions. To detect cultural differences in communication strategies and patterns,

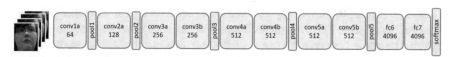

Fig. 7.2 The details of C3D architectures

the games were played in eight different locales across the world. In the following we present details on dataset collection in terms of different locations, participants, procedure, game play and measurements.

Participants

There are 693 participants recruited via email, message boards and advertisements from public universities in the Southwestern US (9 games; $n = 59$), Western US (11 games; $n = 67$), Northeastern US (10 games; $n = 74$), Israel (10 games; $n = 71$), Singapore (12 games; $n = 84$), Fiji (14 games, $n = 106$), Hong Kong (15 games, $n = 115$), and Zambia (15 games, $n = 117$). The sample was 59% female, and was ethnically diverse, with the biggest groups being Asian (38%) or white, non-Hispanic (18%). Nationalities represented 41 different countries. Participants were required to be proficient English speakers. Each game was approximately 2 hours long.

Procedure

Participants enrolled using an online scheduling system. Groups ranged from five to eight participants. Prior to arrival at the site, participants completed consent forms, cultural measures, and demographic questions. Upon arrival, participants were randomly assigned to one of eight computer equipped with a desk, a computer tablet with a built-in webcam, and a chair. Participants were informed that they would be filmed by the cameras.

Each group had a facilitator who explained the rules of the game. Interaction began with an ice-breaker activity, after which players rated each other on scales meant to capture baseline perceptions of dominance, composure, and trustworthiness. Participants took part in the game for an hour, during which they played between three and eight rounds. After the second, fourth, and sixth rounds, and at the end of the game, participants again completed ratings of one another and identified who they thought were the spies. Participants were paid for participating and received additional financial incentives for performing well in the game.

Game Play

Similar to (Zhou 2013), we adapted a version of the Mafia game that closely resembles the board game The Resistance. We pilot tested several versions of the game to ensure the game best met the needs of the research questions. Players were randomly and secretly assigned to play deceivers (called "Spies"), or truth tellers (called "Villagers"). In games of five or six players, two were assigned to be Spies, and in games of seven or eight players, three were assigned to be Spies.

The goal of Villagers was to remove Spies from their community; the goal of Spies was to undermine the missions of the Villagers. Spies were aware of who the other Spies were, but Villagers did not. Villagers had to depend on shared information to deduce the other players' identities within the game.

Players completed a series of "missions" by forming teams of varying size. At the beginning of each round, players elected a leader, who then chose other players for these missions based on who they thought would help them win the game. All players voted to approve or reject the team leader, then voted on the leader's proposed team. Players voted secretly on their computer and publicly by a show of hands. Facilitators would announce if there was a discrepancy in public and private votes, thus informing participants when deception occurred. Players chosen by the leader to go on a mission team secretly voted for the mission to succeed or fail. Villagers won rounds by figuring out who the spies were and excluding them from the mission teams. Spies won rounds by causing mission failures. The ultimate winner of the game (Spies or Villagers) was determined by which team won the most rounds. Additionally, players won monetary rewards by being voted as leader or winning the game.

Measures

We design several measurements for monitoring the game play, including Game Outcome, Trust, Dominance and Previous Game Experience.

Game Outcome   In (Zhou 2013) Mafia study, they regard the deception detection success as the truth-tellers winning the game (i.e., if the truth-tellers win, they must have accurately detected deception). Similarly, in this study, game outcome was a dichotomous variable measuring whether or not Spies or Villagers won the game.

Trust   The extent to which participants trusted each of the other players was measured using a single-item repeated measure, which was asked after the icebreaker, and then every even-numbered round during the game. The item read: Please rate how much you trust each player. Are they trustworthy or suspicious? A rating of 5 would mean they seem honest, reliable and truthful and 1 would mean you thought they were dishonest, unreliable and deceitful (1 Not at all to 5 Very much; Mean = 3.29, SD = 1.36). Because participants responded to this item three to five times about each of the other players, we chose to use a single item in order to avoid fatigue.

Dominance   The extent to which participants found other players to be dominant was measured using a single-item repeated measure (after the icebreaker and each of the even numbered rounds). Participants read the following text: Please rate how dominant each player is. Are they active and forceful or passive and quiet? A rating of 5 would mean you thought they were assertive, active, talkative, and persuasive. A score of 1 would mean you thought they were unassertive, passive, quiet and not influential. We got the statistical of Dominance as Mean = 3.28 and SD = .87.

Previous Game Experience Participants' previous experience playing similar games was evaluated after the completion of the post-game measures. Participants indicated that they had or had not played a similar game. In this study, 54.2% said they had not played a similar game before.

Results of Spy Detection

In the experiments, 280 players' videos are collected for Spy/Villager prediction, including 110 spies and 170 villagers, where the players with different culture background are mixed. We segment the video clip of the first game round for training and testing. The total length is 84,000 seconds (1400 min). We randomly select the 10%/20% videos as testing data and the rest as the training, where there is no duplicate players appearing in both training and validation set. For each setting, the experiments are conducted 5-times cross validation and the results are reported in Tables 7.1 and 7.2. Those two tables shows the Spy/Villager predication accuracy with two different frame sampling mechanisms, random sampling and attention guided frame sampling.

Random Sampling During training, given a video file, we random sample 16 frames as the input of C3D and each frame is re-sized to 112×112. The temporal order is kept among the selected frames. The prediction accuracy is shown in Table 7.1, where the C3D model performs better than the random guess of 60% (170/280), in the margin of ~3% and ~5%.

Attention-guided Sampling The crucial difference between our Spy/Village prediction and the use of conventional image or video classification is that even with the label (spy or villager), a human has a hard time to explain the reason why the players are classified as 'Spies'. In most of these cases, spies and villagers have very similar behaviors, which means the data is not discriminative. As the high accuracy of spy prediction is one of our goal, finding where and when the players show the visual cues for 'being a spy' is the goal of our work. As in Table 7.1, the trained deep

Table 7.1 The Spy/Villager prediction accuracy reported on the two different dataset splitting

#Validation/training games classification accuracy	
1/9	65.43(± 0.27)
2/8	62.28(± 0.30)

The training data is randomly sampled without attention knowledge

Table 7.2 The Spy/Villager prediction accuracy reported on the two different dataset splitting

#Validation/training games classification accuracy	
1/9	67.85(± 0.25)
2/8	67.03(± 0.28)

The model is trained with the video frames selected according to the model attention

neural net model demonstrates better performance than random guess, which motivates us to interpret the model so to understand what visual patterns make the model predict the players as 'Spies'. Given the trained C3D model, we apply the proposed harmonic attention mechanism to compute the frame importance via averaging the attention maps. Instead of random sampling the frames in equal probabilities, we sample the frames according to the importance, leading to higher chances to sample the frames with more contribution to Spy/Villager prediction.

We keep all the parameters the same for training models with the two different frame sampling approaches. Table 7.1 shows our classification results when we randomly selected the video clips from the data. Table 7.2 shows the classification results when we retrain the model based on our attention discovered video frames. The results show clearly that the model trained with attention guided frame sampling outperforms the one with random sampling in the notable margins, ~2% and ~4% in testing/training splitting of 1:9 and 2:8 respectively. The performance boosting validates the effectiveness of our attention mechanism to identify the potential frames where spies and villagers show notable discriminative visual signals so that it is easier for training a model with better accuracy. Besides the quantitative results, we also apply the attention map to identify important pixels and visualize them in the next subsection.

Attention and Deception Cues

In Fig. 7.3 we show promising qualitative results on the fact that our attention NN is capable of discovering cues related to what is known from communication theory for deception. In Fig. 7.3 we show some Facial Action Units related to spies extracted from the discovered frames and the respective probabilities, i.e., AUs: 13,20,24,45. The players showing such AUs are more likely to be classified as Spies. According to the communication theory, AUs 20 and 45 are related to deception, which is consistent to our expectation that spies are more willing to lie, but not always. In Fig. 7.4 we show the ability of our network to attend to different cues for a spy and a villager which are also consistent with the current communication theory (DePaulo and Cooper 2003; Burgoon 2014) on deception. Figure 7.4 clearly shows cues and respective pixel probabilities, which are related to deception such as eyes closed, fake smiles, changes in lips. In particular, we show the comparison of model attention between spies and villagers. For example, our approach can attend to small facial movements related to deception like eye blinking in the bottom left case (spies). At the top row (spies and villagers), the model detects the fake and real smile so as to classify the two type of players role, correctly. These initial encouraging results show that we can extract cues and AUs related to communication and deception theory without using a prior known cues. They provide cues which are human interpretable and can be used in many other types of applications.

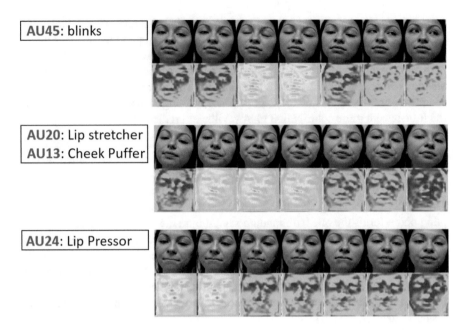

Fig. 7.3 We compare what the model attends to known Action Units which are useful in communication research regarding face and head. The attended facial cues are coded as facial action units (AU). All examples are from spies

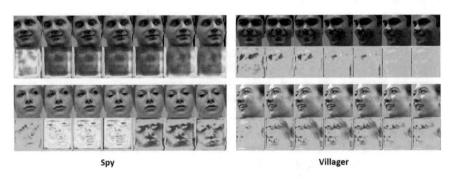

Spy   Villager

Fig. 7.4 What the model attends to for "Spy" *vs.* "Villager". We show the comparison of attention maps between spy and villager. The model can attend to small facial movements related to deception like eye blinking in the bottom left row (spies). And at the top row (spies and villagers), the model detects fake and real smiles so as to classify the two type of players role, correctly

Conclusion

In this paper, we presented a novel attention-based neural network (NN) that discovers through learning in a video sequence the most discriminative frames and related pixel probabilities and AUs that contributed the most to the final class inference of the neural net. We applied our method to facial videos of a variant of the Resistance game collected in various countries where the players assume the roles of deceivers (spies) vs truth-tellers (villagers). We demonstrated for the first time that it is possible to discover the frames and AUs that contributed the most to the NN's class decision on several hours of video testing. The results are consistent with the current communication theory on nonverbal communication and can be used in future studies to discover static and dynamic relationships among cues and AUs currently not known.

Acknowledgement We are grateful to the Army Research Office for funding much of the work reported in this book under Grant W911NF-16-1-0342.

Funding Disclosure This research was sponsored by the Army Research Office and was accomplished under Grant Number W911NF-16-1-0342. The views and conclusions contained in this document are those of the authors and should not be interpreted as representing the official policies, either expressed or implied, of the Army Research Office or the U.S. Government. The U.S. Government is authorized to reproduce and distribute reprints for Government purposes notwithstanding any copyright notation herein.

References

Burgoon, J. K. (2005). Nonverbal measurement of deceit. In V. Manusov (Ed.), *The sourcebook of nonverbal measures: Going beyond words* (pp. 237–250). Hillsdale, NJ: Erlbaum.

Burgoon, J. K. (2015). When is deceptive message production more effortful than truth-telling? a bakers dozen of moderators. In *Frontiers in Psychology, 6*.

Chattopadhay, A., Sarkar, A., Howlader, P., & Balasubramanian, V. N. (2018). Grad-cam++: Generalized gradient-based visual explanations for deep convolutional networks. In *2018 IEEE Winter Conference on Applications of Computer Vision (WACV)* (pp. 839–847). IEEE.

Chaudhry, A., Dokania, P. K., & Torr, P. H. (2017). *Discovering class-specific pixels for weakly-supervised semantic segmentation* (p. 2017). BMVC.

Lindsay, J. J., Malone, B. E., Muhlenbruck, L., Charlton, K., DePaulo, B. M., & Cooper, H. (2003). Cues to deception. In Cues to deception. *Psychological Bulletin, 129*(1), 74–118.

Dosovitskiy, A., & Brox, T. (2016). Inverting visual representations with convolutional networks. In *Proceedings of the IEEE Conference on Computer Vision and Pattern Recognition* (pp. 4829–4837).

Dowdall, J., Shastri, D., Pavlidis, I. T., Frank, M. G., Tsiamyrtzis, P., & Ekman, P. (2007). Imaging facial physiology for the detection of deceit. In *International Journal of Computer Vision, pages 71*, 2, 197–214.

Elkins, A. Burgoon, J. K., Nunamaker, J. F. Jr., Twyman, N. W. (2014). A rigidity detection system for automated credibility assessment. In *Journal of Management Information Systems, 31*, 173–201.

Erhan, D., Bengio, Y., Courville, A., & Vincent, P. (2009). Visualizing higher-layer features of a deep network. *University of Montreal, 1341*(3), 1.

Friesen, W. V., Ekman P. (1975). Unmasking the face. a guide to recognizing emotions from facial clues. In Englewood Cli s, NJ: *Prentice Hall, 71*(2), 197–214.

Guerrero, L., Floyd K., Burgoon, J. K. (2010). Nonverbal communication. In *Allyn Bacon*.

He, K., Zhang, X., Ren, S., & Sun, J. (2016). Deep residual learning for image recognition. In *Proceedings of the IEEE conference on computer vision and pattern recognition* (pp. 770–778).

Hu, J., Shen, L., & Sun, G. (2017). *Squeeze-and-excitation networks.* arXiv preprint arXiv:1709.01507, 7.

Jensen M. L., Kruse J., Meservy T. O., Nunamaker J. F. Jr., Burgoon, J. K. (2007). Deception and intention detection. In Handbooks in Information Systems: National Security, pages Vol 2,193–214.

Jetley, S., Lord, N. A., Lee, N., & Torr, H. P. (2018). *Learn to pay attention.* arXiv preprint arXiv:1804.02391.

Li, K., Wu, Z., Peng, K., Ernst, J., & Fu, Y. (2018). Tell me where to look: Guided attention inference network. CVPR.

Mahendran, A., & Vedaldi, A. (2015). Understanding deep image representations by inverting them. In *Proceedings of the IEEE conference on computer vision and pattern recognition* (pp. 5188–5196).

Metaxas, D., Bourlai T., Elkins A., Burgoon, J. K. (2017). Social signals of deception and dishonesty. In Social signal processing. Cambridge, UK: Cambridge University Press, pages 404–428.

Park, J., Woo, S., Lee, J., & Kweon, I. S. (2018). *Bam: bottleneck attention module.* arXiv preprint arXiv:1807.06514.

Proudfoot, J. G., Wilson D., Schuetzler R., Burgoon, J. K. (2014). Patterns of nonverbal behavior associated with truth and deception: Illustrations from three experiments. In Journal of Nonverbal Behavior, pages 38, 325–354.

Selvaraju, R. R., Cogswell, M., Das, A., Vedantam, R., Parikh, D., & Batra, D. (2017). Grad-cam: Visual explanations from deep networks via gradient-based localization. *ICCV, 1*, 618–626.

Simonyan, K., Vedaldi, A., & Zisserman A. (2013). *Deep inside convolutional networks: Visualising image classification models and saliency maps.* arXiv preprint arXiv:1312.6034.

Springenberg, J. T., Dosovitskiy, A., Brox, T., & Riedmiller, M. (2014). *Striving for simplicity: The all convolutional net.* arXiv preprint arXiv:1412.6806.

Sun, M., Farhadi, A., & Seitz, S. (2014). Ranking domain-specific highlights by analyzing edited videos. In *European conference on computer vision* (pp. 787–802). Springer.

Tran, D., Bourdev, L., Fergus, R., Torresani, L., & Paluri, M. (2015). Learning spatiotemporal features with 3d convolutional networks. In *Proceedings of the IEEE international conference on computer vision* (pp. 4489–4497).

Twyman, N. W., Burgoon, J. K., Nunamaker, J. F., Diller, C. B. R., & Pentland, S. J. (2017). A video-based screening system for automated risk assessment using nuanced facial features. *Journal of Management Information Systems, 34*(4), 970–993.

Wang, F., Jiang, M, Qian, C., Yang, S., Li, C., Zhang, H., Wang X., & Tang, X. (2017). *Residual attention network for image classification.* arXiv preprint arXiv:1704.06904.

Wei, Y., Feng, J., Liang, X., Cheng, M., Zhao, Y., & Yan, S. (2017). Object region mining with adversarial erasing: A simple classification to semantic segmentation approach. *IEEE CVPR, 1*, 3.

Woo, S., Park, J., Lee, J. Y., & Kweon, I. S. (2018). Cbam: Convolutional block attention module. In *Proceedings of the European Conference on Computer Vision (ECCV).*

Yang, H., Wang, B., Lin, S., Wipf, D., Guo, M., & Guo, B. (2015). Unsupervised extraction of video highlights via robust recurrent auto-encoders. In *Proceedings of the IEEE international conference on computer vision* (pp. 4633–4641).

Yao, T., Mei, T., & Rui, Y. (2016). Highlight detection with pairwise deep ranking for first-person video summarization. In *Proceedings of the IEEE conference on computer vision and pattern recognition* (pp. 982–990).

Zafeiriou, S., Pantic, M., Burgoon, J. K., Elkins, A. (2015) Unobtrusive deception detection. In In R. Calvo, S. K. DMello, J. Gratch, A. Kappas (Eds.), The Oxford handbook of affective computing. UK: Oxford University Press.

Zeiler, M. D., & Fergus, R. (2014). Visualizing and understanding convolutional networks. In *European conference on computer vision* (pp. 818–833). Springer.

Zhang, D., Sung, Y., Zhou, L. (2013). The e ects of group factors on deception detection performance. In *Small Group Research, 44*, 272–297.

Zhang, X., Wei, Y., Feng, J., Yang, Y., & Huang, T. (2018). Adversarial complementary learning for weakly supervised object localization. In *IEEE CVPR*.

Zhou, B., Khosla, A., Lapedriza, A., Oliva, A., & Torralba, A. (2016). Learning deep features for discriminative localization. In *Proceedings of the IEEE Conference on Computer Vision and Pattern Recognition* (pp. 2921–2929).

Chapter 8
Iterative Collective Classification for Visual Focus of Attention Prediction

Chongyang Bai, Srijan Kumar, Jure Leskovec, Miriam Metzger,
Jay F. Nunamaker, and V. S. Subrahmanian

Introduction & Motivation

Given a short video clip (1/3rd sec) of a group G of people, for any person $P \in G$, our goal is to build a system to predict the Visual Focus of Attention (VFOA) of P, i.e., what/who P looks at. VFOA is a crucial piece of information when studying various types of social behaviors in a group, e.g., dominance (Bai et al. 2019a), and trust (Knapp et al. 2013).

Figure 8.1a shows some challenges to solve the problem. First, a person's VFOA can change rapidly even within 2 seconds. In Fig. 8.1a, the lady was looking at person 6, person 1, person 1 again, and then person 7. Second, during multi-person interactions, any person can speak at any time, which can heavily influence the VFOA of all people since people's gaze is often directed at speakers. Therefore, our system needs to capture the rapid change of VFOA. Third, nonverbal behaviors (e.g. eye and head movements) of people may also influence the VFOA. As shown in Fig. 8.1a, anyone who speaks or makes obvious body gestures can quickly attract the subject's attention and become her VFOA. Alternatively, people's VFOAs are not mutually independent and may influence one another – for instance, they might all look at the speaker or a person who raises their hand. In short, predicting the

C. Bai · V. S. Subrahmanian (✉)
Dartmouth College, Hanover, NH, USA
e-mail: chongyang.bai.gr@dartmouth.edu; vs@dartmouth.edu

S. Kumar · J. Leskovec
Stanford University, Stanford, CA, USA

M. Metzger
University of California, Santa Barbara, Santa Barbara, CA, USA

J. F. Nunamaker
University of Arizona, Tucson, AZ, USA

© Springer Nature Switzerland AG 2021   139
V. S. Subrahmanian et al. (eds.), *Detecting Trust and Deception in Group Interaction*, Terrorism, Security, and Computation,
https://doi.org/10.1007/978-3-030-54383-9_8

Person 3

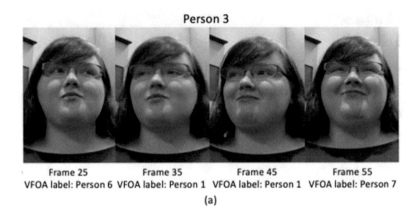

Frame 25   Frame 35   Frame 45   Frame 55
VFOA label: Person 6 VFOA label: Person 1 VFOA label: Person 1 VFOA label: Person 7

(a)

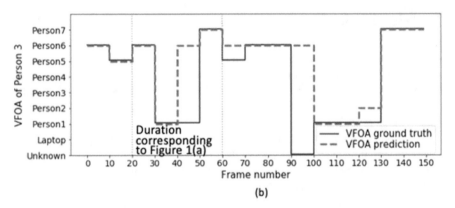

(b)

Fig. 8.1 (**a**) An example of a person 3's VFOA in 4 frames within 4/3 second (40 frames). (**b**) Person 3's VFOA label and prediction made by the ICAF system

VFOA of a single person requires considering the behavior of all people in the group at the sub-second level.

To address these challenges, we propose a system called ICAF (which stands for Iterative Collective Attention Focus) which: (i) predicts the VFOA at the 1/3 second level, which previous studies have shown to be the smallest duration that humans need to focus their attention visually (Rayner 2009), (ii) leverages collective classification (Sen et al. 2008; Kong et al. 2012) to jointly predict VFOAs of all people simultaneously rather than predicting each person's VFOA independently – this uses the intuition that where person P is looking influences, and is influenced by, where others are looking, and (iii) ICAF refines the predictions iteratively by a multi-layer architecture. (iv) ICAF uses the temporal consistency of VFOA, e.g. the conditional probability that P is looking at Q, given that she was looking at Q in the previous 1/3 sec. To the best of our knowledge, we are the first to use where others are currently looking to make simultaneous predictions of VFOA.

To make our system general enough to apply to videos that have never been seen before, we create a lightly supervised label generation method that uses the speaker label to approximate the VFOA label. We call this system LightICAF. LightICAF has comparable performance to ICAF, showing the potential of using ICAF for previously unseen videos. The detailed description of the LightICAF algorithm is presented in (Bai et al. 2019b). The system demo and predicted VFOA networks are available at https://dsaildartmouth.github.io/SCAN.pdf.

System Architecture

We describe our ICAF system in this section. Figure 8.2 shows the ICAF architecture. First, the system extracts the features from both training videos and testing videos. Second, the system generates labels from training videos. Third, the ICAF model is trained and saved for inference – we will describe the model later. Finally, given a new video, the system predicts the VFOA network from the extracted features and the trained model.

Feature Extraction

We extract two sets of features from the clips: face attributes and speaking probability features. For facial attributes, we extract the person's head pose angles and eye gaze vectors using OpenFace (Baltrusaitis et al. 2018) since the tablet cameras can capture close-up face of each person. We build a model to predict the speaking probabilities from facial key points provided by OpenFace.

Speaking Prediction

We predict if a person is speaking at each 1/3 second as follows: first, we get 2-dimensional lip contour points at each frame from OpenFace and normalize them by their bounding box to avoid the influence of head movement. Second, we

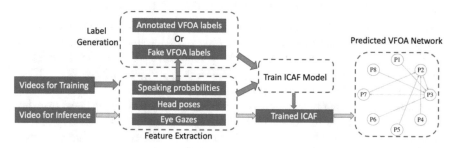

Fig. 8.2 Architecture of our system

compute the gradient of each normalized point's positions over time in order to capture mouth movement, and aggregate the gradients of points at each frame as a frame feature vector. Third, we concatenate the feature vectors of frames around each frame in a window, thus forming a sliding window over time. We use the concatenated feature vectors to train a general speaking detection model, **SP**.

Label Generation

We generate labels in two different ways: supervised annotation, and light supervised approximation. The labels are: other people in the group and the frontal laptop of this person. In the supervised approach, a human expert annotates a person's VFOA in a clip of 1/3 second based on her frontal video and the global-view video of the whole group. In the light supervised approach, we use the single speaker as other people's VFOA. The intuition is that people are highly likely to look at the person who is speaking if there is a single speaker (Stiefelhagen et al. 2002). Therefore, we detect clips where only one person in the group is speaking by using the pre-trained speaker prediction model **SP** described in Section "Feature Extraction". In order to reduce false positives, we smooth the prediction probabilities by a window of 10 frames, and remove clips that are less than 5 seconds. Specifically, for a segment where only person P_i is speaking, we assign i as the training label for all other people and the model is trained with it. To get clips with 'frontal laptop' labels, we choose clips that longer than 10 seconds where no one is predicted as speaking.

ICAF: Iterative Collective Classification

In this subsection, we describe ICAF, which considers inter-person and temporal relationship to collectively classify the VFOA of the whole group.

Suppose there are k people in a group, and suppose $f_{i,t}$ is feature vector that we extracted for person p_i at time t. $f_{i,t}$ is the concatenation of the head pose vector, the eye gaze vector and speaking probability vector $\vec{s} = \left(s_1,\ldots,s_{i-1},0,s_{i+1},\ldots,s_k\right)$. We require that $s_i = 0$ as P_i's speaking activity doesn't influence her VFOA. Suppose we use C_i to denote the VFOA classifier of person P_i. ICAF trains C_i for each person P_i. At time t, $v_{i,t}$ is the probability of P_i looking at person P_j (or the frontal tablet) for each j. The ground truth of person P_i's VFOA is $y_{i,t}$. Figure 8.3 shows the ICAF architecture as an L-layer network. The model builds layers 1, …, L and iteratively uses the output of other people's classifiers as input (shown in dotted lines) from last layer. As shown in dashed lines, each classifier additionally uses the output from time $t - 1$ as input (only person 1 is showed for simplicity). The final prediction vectors are $v_{i,t}^{(L)}$. Algorithm 1 shows the overall algorithm of ICAF.

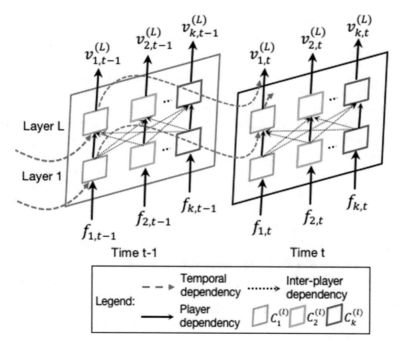

Fig. 8.3 ICAF architecture. Best viewed in color

Algorithm 1: *ICAF* MODEL

Input   : Raw features $f_{i,t}$ $\forall i \in [1, \ldots k], t \in [1, \ldots T]$,
            Number of layers L.

Output: Predictions $v_{i,t}^{(L)}$ of all people i at all times t

1 $v_{i,0}^{(l)} = (\frac{1}{k+1}, \frac{1}{k+1}, \ldots, \frac{1}{k+1})$
2 $v_{i,t}^{(0)} = C_i^{(0)}(f_{i,t})$
3 **for** $t \in [1, \ldots T]$ **do**
   　/* Operate on every time step t */
4 　**for** $l \in [1, \ldots L]$ **do**
   　　/* Process every layer l */
5 　　**for** $i \in [1, \ldots k]$ **do**
   　　　/* Update person P_i */
6 　　　$S(V) = \sum_{j \in \{1, \ldots k\} - \{i\}} v_{j,t}^{(l-1)}$
7 　　　$v_{i,t}^{(l)} = C_i^{(l)}(f_{i,t}, v_{i,t}^{(l-1)}, v_{i,t-1}^{(l-1)}, S(V))$
8 　　**end**
   　　/* Make prediction and save $C_i^{(l)}$
   　　 */
9 　**end**
10 **end**
11 **return** $v_{i,t}^{(L)}$ $\forall i \in [1, \ldots k], t \in [1, \ldots T]$

Inter-Person Dependencies In multi-person discussions, the behavior of people is highly correlated – e.g. people are likely to look at the speaker (Ba and Odobez 2011). This kind of mutual influence can be used to improve the prediction accuracy.

We incorporate the person-to-person influence by adding explicit connections between their classifiers (lines 4–8 in Algorithm 1). We build multiple layers to feed the output of layer $l - 1$ as the input to layer l. Thus, the input to person P_i's model $C_i^{(l)}$ at layer l is its output from layer $l - 1$ and an aggregation of the outputs from other people's classifiers from layer $l - 1$. We use the summation $S(V)$ to aggregate, then input it to the classifier at the next layer (lines 6–7 in Algorithm 1).

For layer 1, let $\vec{v}_{i,t}^{(0)} = C_i^{(0)}\left(\vec{f}_{i,t}\right)$, where $C_i^{(0)}$ is the classifier trained by only features (head pose, eye gaze, speaking probabilities) of P_i, separately. This training is the warm-up which results in more robust iterations.

Temporal Consistency The VFOA of a person at time t is related to her VFOA at time $t - 1$. For example, it is most likely that a person will look at targets around the current target in the short future. The temporal consistency component adds the predictions from the previous time point for a person to the input of classifier at the current time point for the same person. Specifically, the output $\vec{v}_{i,t-1}^{(l-1)}$ is an input to $C_i^{(l)}$. This is shown using the dashed lines in Fig. 8.3 and in line 7 in Algorithm 1. As initialization for each layer l, we require that $\vec{v}_{i,0}^{(l)}$ be a uniform probability distribution for VFOA targets.

The final formulation is shown in Fig. 8.4. Overall, ICAF uses features from faces, temporal features, and inter-person dependencies to jointly predict the visual focus of attention of all people.

$$\vec{v}_{i,t}^{(l)} = C_i^{(l)}(\underbrace{\vec{f}_{i,t}}_{\text{Raw input}}, \underbrace{v_{i,t}^{(l-1)}}_{\text{Participant input}}, \underbrace{v_{i,t-1}^{(l-1)}}_{\text{Temporal input}}, \rho(\underbrace{\sum_{j\in\{1,\dots k\}-\{i\}} \phi(v_{j,t}^{(l-1)})}_{\text{Inter-participant input}})))$$

Fig. 8.4 Final formulation of ICAF to output $\vec{v}_{i,t}^{(l)}$ of person i at time t on layer l

VFOA Network Prediction

Once the ICAF models of all layers for all people are trained, we are ready to apply them to the group as a whole in order to predict the VFOA network, i.e. the network whose nodes are people and whose edges denote "who looks at who". Given a clip, we extract features from all frames and apply the model in Algorithm 1. We then average the VFOA probabilities of frames in every 1/3 second for each person, which results in the VFOA probabilities for every 1/3 second of the group. For each person, we get the target VFOA with the maximum probability of all her VFOA probabilities. We define the people in the group as nodes in the VFOA network, and the weighted edges as the obtained probabilities. In the visualization of Fig. 8.2, the thicker edge denotes larger weights, i.e. more confident prediction.

System Results

In this section, we describe the results returned by our system. In particular, the ICAF system shows the dynamic facial attributes of the person being considered, as well as predicted probabilities of the people that the subject is looking at, in addition to the videos.

The ICAF system takes the frontal videos of a group of people as input. Users can choose videos to load to the system. The following components are outputs generated by the system.

Facial Features For each person, our system returns the head pose vector, the eye gaze vectors, and the facial key points. The results are extracted by OpenFace. Figure 8.5 visualizes a snapshot of one player. The head pose is showed as the projected blue cube, the 2 eye gaze vectors are showed as the green segments, and the facial key points are showed in red.

Result Panel For each person, our system returns the speaking and VFOA predictions. As shown in Fig. 8.6, at every 1/3 second, it returns the probability of a person speaking, which is highlighted as green when larger than 0.5. Moreover, it shows the probabilities of looking at other people and looking at her frontal laptop. The highest VFOA probability is also highlighted.

Predicted VFOA Networks At each 1/3 second, the network has people in the group as vertices, and people's VFOA as edges. A directed edge is defined from the vertex of person to it of her VFOA target. Note that the self-loop of a vertex indicates that the person looks at her laptop. The larger the probability is, the thicker the edge will be. Figure 8.7 shows 2 examples of the predicted VFOA networks. In the right example of Fig. 8.7, it is clear that all other people look at person 4, while person 4 looks at person 1. In the video, person 4 was speaking to person 1 so everyone else was looking at her. In the left example, person 2 looks at her own laptop,

Fig. 8.5 Facial features of
person 6 at 1/3 second of
the video

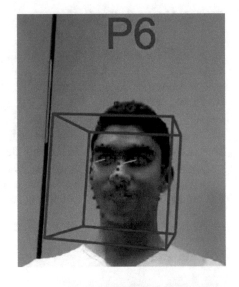

Fig. 8.6 The result panel
with probabilities and
predictions

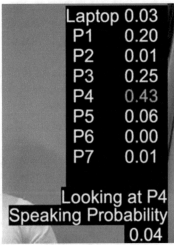

and the system predicts person 1 and 4 gazing at each other with high probability –
note that the edges in question are thicker than other edges.

Saved Probabilities Files In addition to the visualized result, our system also
saves all the predicted speaking and VFOA probabilities as files. We have released
the generated VFOA probabilities at http://snap.stanford.edu/data/comm-f2f-Resistance.html.

Overall, Fig. 8.8 puts all results together from a 1/3 second, resulting one frame
of the visualized videos by our system.

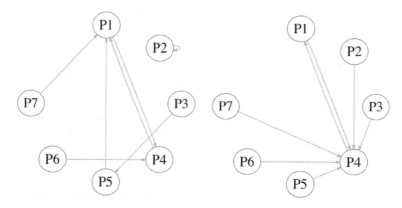

Fig. 8.7 Two examples of the VFOA networks in a same video

(a) Facial features (b) Result panel   (c) Predicted VFOA network

Fig. 8.8 System Return

Experiments

We conduct through experiments on our Resistance dataset and the AMI (McCowan et al. 2005) dataset to show:

1. For the task of predicting VFOA in the next 1/3 second (i.e., 10 frames), ICAF outperforms all baselines by 1.3% ($p < 0.05$ by two-sample t-test).
2. For the task of making longer figure predictions, ICAF significantly outperforms the highest baseline by up to 5% ($p < 0.05$).
3. Both collective and temporal components boost the performance of ICAF significantly.
4. The lightly supervised prediction gets comparable results to supervised prediction in both ICAF and baselines.

Baselines We compare with three sets of baselines that use head pose vector (H), eye gaze vector (E), and speaking probability vector(S) as features. The first set of baselines are (Ba and Odobez 2009; Ba and Odobez 2011; Massé et al. 2017). Specifically, GMM(H), GMM(H,E) use Gaussian Mixture Model with parameters from each individual (Ba and Odobez 2009). HMM(H), HMM(H,E) uses Hidden Markov Model (Ba and Odobez 2009). DBN(H,S), DBN(H,E,S) uses Dynamic Bayesian Network (DBN) incorporating conversational dynamics. G-DBN uses DBN to track VFOAs and eye gaze simultaneously with people's global head poses as inputs (Massé et al. 2017). Further, we invented two more sets of baselines. The second set of baselines trains one general classifier GC for all people. To reduce the confusion of different gaze behaviors caused by positions (e.g. a right person turning head left and a left person turning head right can lead to the same VFOA target), we add the person index to the feature vector (Ba and Odobez 2011). The last set of baselines trains a person-specific classifier PC for each person (Asteriadis et al. 2014).

Experimental Setting For speaking prediction, we set the sliding window size to 1 sec (30 frames) and train a Random Forest classifier **SP**. The positive training samples are people's self-introductions at the start of the game, while the negative ones are the clips of people other than the introducer at that same point in time. Note that these clips are not from any data used in ICAF model. We evaluate according the temporal order of data. Specifically, we train the model on the first T data points and test on the $T + k^{th}$ data point (each data point has 10 frames). T is varied from 96.3% to 99.9%. The video for each game is divided into three parts: an introduction round and two discussion rounds. We train the models using the clips from the speaker's self-introduction, and evaluate them temporally using clips from the two rounds of discussion. Both training and testing are at the frame level. Frame VFOA probabilities are averaged over 10 frames as probabilities at each 10-frame clip. We experiment ICAF with 4 basic classifiers: Random Forest (RF), Logistic Regression (LR), Linear SVM (LINSVM) and Gaussian Naive Bayes (NB). In all cases, ICAF has 3 layers. We report accuracy for all experiments.

Experiment 1: Next VFOA Prediction We compare ICAF with all baselines using all three sets of features. All models are trained on the first T data points and then used to predict the $T + 1$ data point. Table 8.1 shows the results. (i) Person-specific baselines are better than the corresponding general classifier baselines using the same set of features. Specifically, PC(H,E,S) performs at least 6.2% better than GC(H,E,S). (ii) ICAF performs 1.3–11.2% better than all baseline models.

Experiment 2: Longer-Future Predictions We next evaluate the robustness of ICAF by predicting the $T + k^{th}$ data point while training only till the Tth data point. We vary k from 1 to 10, meaning that we predict who a person will look at between 0.3 and 3.3 seconds into the future. Figure 8.9 shows the result. ICAF outperforms the best baseline by up to 5%. Moreover, ICAF's prediction accuracy varies only 7.5% over k, which is robust in the longer-term future.

Experiment 3: Contribution of Collective Classification Figure 8.10 compares the results of ICAF with and without the temporal and collective classification components. Note that ICAF without both components is equivalent to PC(H,E,S). Each of the components boost the performance of ICAF from 0.2% to 5.3% w.r.t. all base classifiers. Additionally, adding collective classification improves performance more than the temporal component alone.

Experiment 4: Comparison with Different Features We compare the performance of different features on all methods. RF is used as the (base). Table 8.2 shows the results for next VFOA prediction. (i) for all models, eye gaze features E boost the predictions. (ii), speaking features S boost all models except for GC. Third, using features with E or S, ICAF is better than all baselines.

Experiment 5: Comparison Between Different Base Classifiers Here we explore performance of ICAF with different kinds of base classifiers: Random Forest, Logistic Regression, and Naïve Bayes. In Fig. 8.11 we compare ICAF with GC and PC. The colored texts show the results for $k = 1$, where ICAF outperforms the corresponding best baseline by 1.3–11%. For $k > 1$, it outperforms the best baseline by up to 5% with RF, 12% with LR, and 4% with NB.

AMI Corpus Experiments We also experimented on the AMI meeting corpus (McCowan et al. 2005). In this dataset, 8 meetings are dynamic, where people sit around a table and up to 1 person moves to the whiteboard/screen to present. 4 meetings are static, where all people remain seated. We followed the leave-one-out protocol as in (Ba and Odobez 2011) and compare frame-based accuracy. *Since the 4 seats over all meetings are fixed, we train seat-specific classifiers in* ICAF. Table 8.3 shows that ICAF outperforms (Ba and Odobez 2011) in both static and dynamic meetings.

LightICAF VFOA Prediction Figure 8.12 shows the results for all methods. Since the training labels come from speaking labels, we remove speaking probability features. Compared to random prediction of 14.4%, lightICAF generates 41.2–54.7% results. In addition, LightICAF is better than the baselines. As a point of comparison, we also train the models using the human-annotated labels in the self-introduction rounds. LightICAF gets similar results as in the case of supervised prediction, which suggests that is effective and generalizable to videos that the system has never seen before.

Future Work

We discuss three potential future directions here:

Table 8.1 Experiment 1: Next VFOA Prediction

GMM(H,E)	HMM(H,E)	DBN(H,E,S)	G-DBN	GC(H,E,S)	PC(H,E,S)	ICAF
0.716	0.770	0.800	0.782	0.756	0.818	**0.831**

Fig. 8.9 Experiment 2:
Longer-Future Predictions

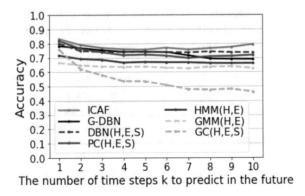

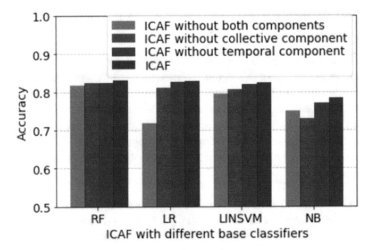

Fig. 8.10 Experiment 3: Contribution of collective classification

1. ICAF iteratively trains a model by adding the previous predictions to the input. We would like to modify it as a collective of multi-layer RNNs. The current layer of each person's RNN builds on top of last layers of all people's RNNs. This would allow the ICAF system to be trained from end to end and may potential lead to more accurate predictions.
2. In the light supervised label generation method, we can add a filter to remove the noisy labels by clustering the head pose and eye gazes vectors of each VFOA class.
3. Our extracted VFOA networks can be further used to annotate the verbal interactions, and analyze social behaviors in a group of people, such as dominance, who supports/opposes who, and who likes/dislikes who.

Table 8.2 Experiment 4: Comparison between different features

Model	GMM	HMM	DBN	GC	PC	ICAF
H	0.525	0.623	–	**0.719**	0.716	0.718
H,E	0.716	0.770	–	0.799	0.805	**0.811**
H,S	–	–	0.665	0.731	0.771	**0.784**
H,E,S	–	–	0.800	0.756	0.818	**0.831**

Prediction Accuracy

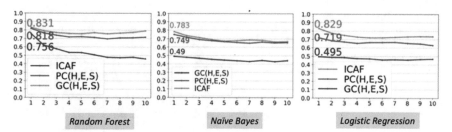

Fig. 8.11 Experiment 5: Comparison between different (base) classifiers

Conclusion

We propose a system to predict the visual focus of attention (VFOA) of people in videos. We showed that explicitly incorporating inter-person dependencies and temporal consistency are crucial if we wish to accurately predict VFOA both in the short-term future and in the longer term. The ICAF model is, therefore, able to overcome the challenges of rapidly changing VFOA, high dynamics of the discussion, and person-person inter-dependencies. Moreover, the lightly supervised ICAF is crucial in making the model general to unseen videos. This opens doors to new research in efficient extraction of interaction networks from videos without any training labels.

Acknowledgement We are grateful to the Army Research Office for funding much of the work reported in this book under Grant W911NF-16-1-0342.

Funding Disclosure This research was sponsored by the Army Research Office and was accomplished under Grant Number W911NF-16-1-0342. The views and conclusions contained in this document are those of the authors and should not be interpreted as representing the official policies, either expressed or implied, of the Army Research Office or the U.S. Government. The U.S. Government is authorized to reproduce and distribute reprints for Government purposes notwithstanding any copyright notation herein.

Table 8.3 AMI corpus experiments		
Model	Static meetings	Dynamic meetings
Ba and Odobez (2011)	0.556	0.520
ICAF	0.568	0.538

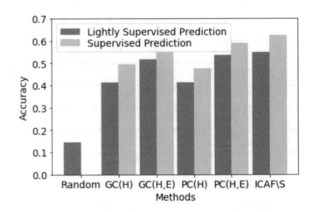

Fig. 8.12 Lightly supervised predictions (in dark green) and supervised predictions (in light green): 'Random' denotes random prediction accuracy, and ICAF\S denotes ICAF without speaking feature

References

Asteriadis, S., Karpouzis, K., & Kollias, S. (2014). Visual focus of attention in non-calibrated environments using gaze estimation. *International Journal of Computer Vision, 107*(3), 293–316.

Ba, S. O., & Odobez, J. M. (2009). Recognizing visual focus of attention from head pose in natural meetings. *IEEE Transactions on Systems, Man, and Cybernetics, Part B (Cybernetics), 39*(1), 16–33.

Ba, S. O., & Odobez, J. M. (2011). Multiperson visual focus of attention from head pose and meeting contextual cues. *IEEE Transactions on Pattern Analysis and Machine Intelligence, 33*(1), 101–116.

Bai, C., Bolonkin, M., Kumar, S., Leskovec, J., Dunbar, N., Burgoon, J., & Subrahmanian, V. S. (2019a). *Predicting dominance in multi-person videos*, International Joint Conference on Artificial Intelligence (IJCAI).

Bai, C., Kumar, S., Leskovec, J., Metzger M., Nunamaker, J. F., & Subrahmanian, V. S. (2019b). *Predicting the visual focus of attention in multi-person discussion videos*. International joint conference on artificial intelligence (IJCAI).

Baltrusaitis, T., Zadeh, A., Lim, Y. C., & Morency, L. P. (2018). Openface 2.0: Facial behavior analysis toolkit. In *2018 13th IEEE international conference on automatic face gesture recognition (FG 2018)* (pp. 59–66).

Knapp, M. L., Hall, J. A., & Horgan, T. G. (2013). *Nonverbal communication in human interaction*. Cengage Learning; 8th Edition (January 1, 2013).

Kong, X., Yu, P. S., Ding, Y., & Wild, D. J. (2012). Meta path-based collective classification in heterogeneous information networks. In *Proceedings of the 21st ACM international conference on information and knowledge management* (pp. 1567–1571). ACM.

Masse,́. B., Ba, S., & Horaud, R. (2017). Tracking gaze and visual focus of attention of people involved in social interaction. *IEEE Transactions on Pattern Analysis and Machine Intelligence, 40*, 2711.

McCowan, I., Carletta, J., Kraaij, W., Ashby, S., Bourban, S., Flynn, M., Guillemot, M., Hain, T., Kadlec, J., & Karaiskos, V. (2005). The ami meeting corpus. In *Proceedings of the 5th international conference on methods and techniques in behavioral research* (Vol. 88, p. 100).

Rayner, K. (2009). Eye movements and attention in reading, scene perception, and visual search. *The Quarterly Journal of Experimental Psychology, 62*(8), 1457–1506.

Sen, P., Namata, G., Bilgic, M., Getoor, L., Galligher, B., & EliassiRad, T. (2008). Collective classification in network data. *AI Magazine, 29*(3), 93.

Stiefelhagen, R., Yang, J., & Waibel, A. (2002). Modeling focus of attention for meeting indexing based on multiple cues. *IEEE Transactions on Neural Networks, 13*(4), 928–938.

Part III
SCAN Project Foundations: Preceding Empirical Investigations of Deception

Chapter 9
Effects of Modality Interactivity and Deception on Communication Quality and Task Performance

Joel Helquist, Karl Wiers, and Judee K. Burgoon

Introduction

Globalization has ushered in frequent reliance on computer-mediated communication (CMC), distributed teams, and a variety of new social media. Not all of these new media present the end user with the same ability to communicate in an effective and efficient way or to detect the presence of "bad actors." Although past research has compared CMC modalities to face-to-face (FtF) interactions in resolving ambiguity, decision-making, and social judgments of teammates (e.g., Daft and Lengel 1986; George et al. 2016; Hiltz 1988; Rice 1992; Vickery et al. 2004; Walther 1992, 1996), little research has examined how communication itself is influenced by the modality being used. It may be that the effect of CMC on group performance is impacted by the qualities of communication that are afforded, fostered, or inhibited within a given communication modality. The quality of communication in a group may directly affect the ultimate degree of success of the group. Especially important for the current goal of detecting deceit, the level of interactivity afforded by different modalities may facilitate or impair the ability of team members to detect ulterior motives and duplicity. Failure to do so may enable those with malintent to sabotage a group. Previous research has shown that as much as one-third of daily communication includes some form of deception such as concealment, omissions, exaggeration, equivocation, or outright falsification of information (Buller et al. 1996; Ekman 1996) (although a few prolific liars may be responsible for a disproportionate amount of outright lying; Serota et al. 2010). If CMC enables such

J. Helquist
Utah Valley University, Orem, UT, USA
e-mail: joelh@uvu.edu

K. Wiers · J. K. Burgoon (✉)
University of Arizona, Tucson, AZ, USA
e-mail: judee@email.arizona.edu

© Springer Nature Switzerland AG 2021
V. S. Subrahmanian et al. (eds.), *Detecting Trust and Deception in Group Interaction*, Terrorism, Security, and Computation,
https://doi.org/10.1007/978-3-030-54383-9_9

misrepresentations or disinformation to go undetected, deceivers will be able to adversely affect group performance. However, if deceit is noticeable during a group's interactions, communication qualities may register its presence.

Background

Principle of Interactivity

Interactivity is a term that has been applied to both the nature of the communication that occurs within a given modality (the communication process) and the affordances (structural characteristics) of the modality itself. Our view is that "interactive" refers to exchanges between sender and receiver in which messages are interdependent, contingent, and nonrecursive (Burgoon et al. 2000). The messages between the sender and receiver build upon each other, with each successive communicative exchange building on earlier exchanges. This interactivity is affected by the characteristics of the communication modality. Interactive communication modalities are ones that foster such interaction through their structural affordances. For example, a conversation using an instant messaging modality hinges on the ability of each participant to receive each message, as successive conversation is based on shared understanding created during the exchange. Communication modalities vary in the extent to which they afford, foster, or inhibit interactive communication processes.

The principle of interactivity states that human communication processes and outcomes vary systematically with the degree of interactivity afforded by the communication modality (Burgoon et al. 2002a). Viewed as a systems model, a communication modality's structural affordances or characteristics are system inputs that can facilitate, inhibit, or preclude the quality of interaction that ensues and the resulting output productivity, including credibility assessments.

The interactivity of a given modality is by itself neither inherently positive nor negative. The impact of interactivity on a given group's performance is dependent on many factors, including such things as the nature of the task, the goal of the interaction, and the composition of the participants (Burgoon et al. 2010). According to media synchronicity theory (Dennis et al. 1998, 2008; Dennis and Valacich 1999), group communication tasks can be classified into one of two general communication processes: conveyance and convergence. The objective of conveyance tasks is to exchange information among the participants so that each team member can formulate an understanding of the information. Convergence tasks are concerned with developing a shared understanding and moving toward a consolidated viewpoint. Conveyance and convergence tasks differ significantly from each other, and each may benefit from different levels of interactivity afforded by various communication modalities.

Interactivity of CMC

This research investigates the structural affordances of the following communication modalities: text, audio, and face-to-face (FtF). Four structural properties of communication modalities that determine the level of interactivity are participation, synchronicity, proximity, and richness of social information (Ramirez Jr. and Burgoon 2004).

Participation refers to whether actors send and receive messages from other actors or are relegated to observer status. Online bulletin boards are an example of a noninteractive medium because communication is unidirectional. Lurking in a chat room is likewise noninteractive because lurkers do not interact with chatters. Participation is one of the most obvious defining characteristics of interactivity. As actors move from a participative to an observer role, systematic changes are expected in their level of engagement and sense of connection with other actors. In this research, all participants were active, in groups of three or four.

Synchronicity refers to whether real-time message exchange is possible. Synchronous communication, exemplified by instant messaging or chat, allows for immediate serial message exchange or even simultaneous speech. Synchronous communication enables messages that are strongly interrelated, with successive exchanges constructed directly upon previous utterances in a dynamic, rapid manner. Asynchronous modalities, such as online bulletin boards or email, include time delay between the message exchanges. In asynchronous modalities, each participant is interacting with the CMC interface at different times. Synchronous communication enables participants to query and resolve misunderstandings in an easier, timelier manner than with asynchronous communication media. Research has shown that synchronous communication, compared with asynchronous communication, facilitates more team cohesiveness and involvement, increasing the sense of engagement and team identity (Burgoon et al. 2002b). Moreover, synchronous communication allows for little time to rehearse or edit messages (Dennis et al. 1998, 2008; Dennis and Valacich 1999). In the case of deceptive messages, this lack of rehearsability may undermine the success of the deceiver, as the deceiver may not be able to craft, edit, and polish a deceptive message before it is sent. In the current research, FtF and audio modalities included little if any gaps between messages, resulting in a high level of synchronicity. The text modality afforded the subjects the opportunity to write and edit a message prior to it being sent. Thus, the text modality was not as synchronous as audio and FtF.

Proximity refers to whether the participants are physically co-located or distributed. Proximal groups share the same physical location. As such, these teams have access to all of the nonverbal cues that are present in a FtF interaction. These cues include such features as facial expressions, head and limb movements, gestures, posture, vocal features, conversational distance, and environmental context features. Proximal communication modalities include more than just FtF interactions. Group support systems often provide an environment where team members are proximally located but communicate primarily through a text-based interface. These teams still have access to nonverbal social information and social presence (Biocca et al. 2003). Distributed teams are dispersed across different geographical locations. These teams do not have the accompanying nonverbal cues that provide additional contextual information to group members. Research has shown that interacting in proximity promotes involvement and a sense of connectedness as well as favorable social judgments of team members (Burgoon et al. 2000, 2002a). Teams that work in close proximity feel more unity and team identity. In the current experiment, teams interacting FtF were proximal whereas those interacting via audio or text were distal.

Richness, in this context, refers to the quantity and variety of information and messages that a communication modality enables and has been examined through research surrounding media richness theory. Each communication modality may comprise various communication modes or channels (Carlson and Zmud 1999). Each of the modes or channels can be used to communicate. For example, video-conferencing enables participants to send and receive messages via multiple communication channels such as verbal, vocal, and nonverbal. Media richness theory (Daft and Lengel 1986) argues that media differ in their ability to transmit information and develop shared understanding (Dennis and Kinney 1998; Nardi 2005). Richer media allow for greater language variety, increased number of ways information can be communicated (e.g., tone of voice or nonverbal body language), greater personalization, and more rapid feedback that allows for increased ability to clarify ambiguities (Dennis and Valacich 1999; Dennis et al. 1998). In the current research, the FtF modality is the richest due to its multiple communication channels, followed by audio, and then text-only communication.

Interactivity, as has been described above, possesses similar structural affordances as those outlined by media synchronicity theory (Dennis et al. 1998, 2008; Dennis and Valacich 1999). The theory presents the following characteristics that distinguish CMC: Immediacy of feedback, symbol variety, parallelism (simultaneous conversation), rehearsability, and reprocessability. These six dimensions overlap considerably with the idea of synchronous communication and richness as presented regarding interactivity. The principle of interactivity expands on these dimensions to include how active a participant is (participation) as well as the physical location of the communicators (proximity). Both of these dimensions may play a key role in the quality of a group's communication and how successful the group may be. For example, both of these dimensions determine such things as the level of nonverbal communication that accompanies verbal communication.

In the current experiment, our interest was in examining the impact of each communication modality's interactivity on communication quality and group outcomes. Of the modalities to be examined, FtF is the most interactive as it provides access to all the verbal and nonverbal cues from other actors that provide valuable context and help participants to coordinate and execute the communication exchange. On the other end of the interactivity spectrum, text-based communication provides access only to verbal information that is explicit in the content of the message itself. Table 9.1 presents each modality and its respective interactivity rankings as they are implemented in the current study.

Table 9.1 Comparison of modalities on interactivity

Modality	Participation	Synchronicity	Proximity	Richness
Face-to-face	High	High	High	High
Audio only	High	High	Low	Medium
Text	High	Medium	Low	Low

Previously, FtF was thought to be the ideal environment for fostering trust, involvement, and mutuality among team members. However, research has shown that the visual information afforded in FtF environments may not be necessary for creating involvement and mutuality (cohesiveness) within a team; actors may perform as well as or better under audio communication and may do better detecting deception without the distraction of self-enhancing visual cues (Burgoon et al. 2003, 2008; Stoner 2001). This possibility was explored in the current research.

Impact of Deception

Group work is assumed to be a cooperative endeavor. As such, it can be sabotaged when team members have conflicting goals or ulterior motives that lead to deceptive practices. To the extent that deceivers can capitalize on interactivity to achieve their own ends and to undermine group performance, interactivity may become detrimental rather than beneficial to group performance.

Deception is an omnipresent feature of human interaction. Interpersonal deception theory (IDT) (Buller and Burgoon 1996) posits that deception is a dynamic process whereby the deceiver sends messages that are received and processed by the receiver. The receiver uses all information channels available, both verbal and nonverbal, to process and interpret the deceiver's message. Thus, the deceiver may use any or all of the channels to successfully execute the deception. The deceptive process continues with the receiver reacting to the messages that were received and exhibiting a set of response messages to the deceiver. The deceiver reacts to the receiver's cues and adjusts the deceptive messages accordingly. IDT posits that the deception process is dynamic and dependent on information cues sent and processed by both the deceiver and the receiver. The deceiver adjusts and controls the information to adapt to the receiver's verbal and nonverbal cues.

Interactivity affects the ability of deceivers to adjust and adapt their deceptive messages to the feedback from the receiver. In "leaner" environments that foster less interactivity, the deceiver is not provided with as many cues regarding how the receiver is receiving and interpreting the deceptive messages. The deceiver is not afforded as many opportunities to adjust and create a more compelling deceptive exchange. We expected that more interactive modalities avail deceivers more cues to monitor, modify, and repair the deceptive messages.

Communication Quality Model

The effects of interactivity on group outcomes can be better understood by considering the qualities of communication that accompany group interaction (Stoner 2001). Stoner (2001) and Burgoon et al. (2003) identified thirteen dimensions of communication quality that principal components factor analysis suggested could be grouped into three meta-dimensions of relational, interactional, and task communication qualities.

Relational quality concerns the interpersonal relationships among actors. This cluster of dimensions includes involvement, mutuality, similarity, positivity, composure, and persuasiveness of the exchange. This meta-dimension is key to activities like team building.

The *interaction quality* meta-dimension concerns the degree of coordination and ease with which messages are exchanged. This dimension includes interaction coordination, communication appropriateness, expectedness, and message richness. Relational and interaction meta-dimensions capture multifaceted properties of interactivity. The interaction quality meta-dimension, in particular, is a separate construct that reflects the principles of interactivity.

The last meta-dimension, *task quality*, concerns task-related communication byproducts of interactivity. Task quality addresses effectiveness and efficiency in performing the required task, including such items as the team's degree of task orientation (i.e., fraction of the team's total time spent on task-oriented issues), efficiency, and level of critical evaluation of the ideas exchanged. These meta-dimensions of communication quality are not argued to be either exhaustive or mutually exclusive; the meta-dimensions are a means to better understand the extent of interactivity and its impact on the communication process and quality.

The communication qualities are conceived of as subjective indicators of the communication quality during the process; the actors evaluate the communication during the process to determine the kinds of communication that transpired. Figure 9.1, adapted from Stoner (2001), shows the proposed relationships for this study. The structural affordances of the communication modality and deception are posited to impact communication qualities which in turn impact group outcomes.

Hypotheses

The first two hypotheses examine the relationship between the interactivity afforded by each modality and the three communication quality meta-dimensions of relational, interaction, and task quality. Two prior investigations examined these relationships. Stoner (2001) compared audio, audiovisual, and FtF conditions during an extended case analysis task. Results showed that audio ranked the highest on relational and interactional meta-dimensions. Absent from his experiment was the text modality. Burgoon et al. (2003) investigated text, audio, and FtF on various

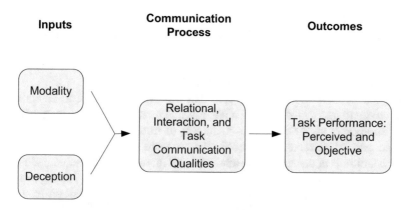

Inputs   **Communication Process**   **Outcomes**

Modality

Relational, Interaction, and Task Communication Qualities

Task Performance: Perceived and Objective

Deception

Fig. 9.1 Model of mediated interaction

relational factors during both a social, get-acquainted game and a task-oriented game. They instead found that FtF earned higher relational quality ratings than audio and text. The current investigation sought to reconcile these mixed findings by including the same three modalities but testing the relationships with a new task. In the experiment reported below, all modalities were synchronous. Audio and text chat were distal while FtF was proximal. Thus, FtF was the most interactive modality. It was hypothesized that text should generate the least engagement and perceived mutuality, and should present the greatest challenge to interaction coordination due to the absence of auditory and visual turn-taking cues. Consequently, it was predicted to fall behind more interactive modalities on relational and interaction quality. At the same time, FtF interaction was predicted to offer no advantage over audio communication due to the offsetting demand for participants to expend cognitive effort to manage social cues and visual appearances. Although Burgoon et al. (2003) showed FtF having higher ratings than audio, those results may have stemmed from the social nature of the task utilized during the experiment. Stoner (2001) found that the audio-only condition received higher task-related ratings than visual formats (videoconferencing and FtF), conceivably due to the reduction in effort directed toward visual social information and greater attention to the task at hand. We reasoned that a task-based scenario would even the playing field between audio and FtF. Given that the current research utilized a structured, task-based experiment, we expected that relational and interaction qualities would earn higher ratings under audio and FtF than text.

H1a: Relational and interaction quality ratings decline as the communication modality changes from audio or FtF to synchronous text.

The second hypothesis was intended to replicate and expand upon Stoner's (2001) finding by including the text modality. Although the text modality can further minimize effort directed toward social information, as a less interactive medium it is also subject to slower information exchange than oral communication. It also presents more opportunities for misunderstanding and less efficient exchanges due to the absence of nonverbal turn-taking cues. The greater difficulty associated with typing rather than speaking may also contribute to less energy devoted to deep analysis, feedback, and evaluation. Text was therefore hypothesized to yield the least, and audio the most, efficient and effective task-oriented communication.

H1b: Task-related communication quality is higher under audio than FtF interaction and higher under FtF than text interaction.

Previous research has demonstrated that deceivers were able to capitalize on interactivity to foster involvement, mutuality, and trust to reduce detection of deception (Burgoon et al. 2003). Under conditions of truth, involvement and mutuality (components of the relational meta-dimension) were similar across modalities, but under conditions of deception, the relational and interactional communication qualities were higher with FtF than mediated conditions and were lowest with text. However, other research (Burgoon et al. 2000) found no significant differences on involvement, mutuality, and similarity for truthful versus deceptive conditions. This research seeks to further expand and clarify these findings by examining the impact of deception on the relational, interactional, and task meta-dimensions.

IDT posits that the presence of deception should negatively impact the deceiver's performance during the early stages of the interaction, leading to communication patterns that deviate from normal behavior. As a caveat, deceivers' repairs of their communication performance over time should mitigate the initial damage, so it was possible that any adverse effects would be transitory.

H2: Compared to truth, deception adversely affects the relational, interactional, and task aspects of communication quality.

Outcomes, or results, can be assessed by team members' perceptions of one another, perceptions of perceived task performance, and actual task performance. Burgoon and colleagues (2000) found that richer modalities resulted in more positive evaluations of team members. Based on the interactivity principle, it is hypothesized that richer media increase the level of interactivity, resulting in higher relational and interaction communication quality. These positive ratings may result in higher measures of perceived task performance, such that the positive ratings from the communication qualities result in a halo effect on perceived task performance.

H3: Higher ratings on relational, interaction, and task communication quality meta-dimensions are associated with more favorable judgments of perceived task performance.

The last hypothesis investigates the relationship between deception and communication quality with objective team performance. Burgoon and colleagues (2000) found partial support for the hypothesis that under deception, performance is negatively associated with involvement, mutuality, and similarity (components of the relational quality meta-dimension). This research seeks to further expand these findings to the three meta-dimensions. It is hypothesized that the greater degrees of positive relational quality, interaction ease and task quality result in a truth bias. Deceivers will capitalize on this truth bias to create a sense of believability and trust. In the case of deceptive information, this truth bias damages the performance of the group due to failure to detect deceptive information.

H4: Relational, interactivity, and task quality meta-dimensions are positively correlated with objective decision quality when there is no deception present but negatively correlated with objective decision quality when deception is present.

It is important to note that the Burgoon experimental design differed in important ways from the current one. It included a two-person team where each member completed a ranking task individually then discussed their separate rankings with their team member. The objective of each team was to attempt to arrive at a consensus regarding a final team ranking. The design required each participant to complete his or her baseline ranking and then justify the ranking to the teammate. The current research design utilized a three-person team with a less structured task that required increased coordination and communication skills, as the team members needed to figure out the best process to complete the task as well as to share relevant information. Both tasks were thus convergence tasks but the current one also required conveyance and more extensive collaboration due to the larger group size and the complexity of the task, thereby increasing the importance of interactivity.

These hypotheses were examined with two different experiments. The first entailed an experimental simulation called BunkerBuster. The second entailed another simulation called StrikeCOM. The second experiment increased sample size and made significant improvements to the simulation. Each data set and their respective results are discussed in turn.

Methods & Results

BunkerBuster Methods

Participants   Participants ($N = 200$) were undergraduate students at a large southwestern university who received course credit for their participation. They were assigned to four-person groups ($n = 50$ groups).

Procedures and Independent Variables   Individuals participated in a board game simulation called BunkerBuster. The goal of the exercise was to have a team work together to find three scud missile launchers that had been hidden within regions of a hypothetical state shown on the board. Each player was assigned a role that had corresponding information assets. These assets included such things as satellites, spies, and unmanned aerial vehicles, each with different capabilities and reliabilities. The team had to coordinate their strategy about how they would search for the scud launchers. The coordination included placement of each asset and sharing of the results returned by the assets. This search planning and information sharing occurred for four search turns. On the fifth turn, the team was required to decide which locations they intended to strike.

Participants were randomly assigned to one of four roles in their group. Each role controlled different intelligence assets (air, space, human intelligence, special ops) that returned information about the likelihood of a searched area containing a missile.

Participants were also randomly assigned to either a deception or truth condition and to one of three communication modalities (FtF, audio, or text). Deception was introduced by instructing one member (space) to provide inaccurate and misleading information so as to lead the group to strike the wrong targets and spare that person's countrymen. To increase motivation, they were told the following:

> *Although most of us typically think that "honesty is the best policy," there are times when being truthful is not in our best interest. In the military, for instance, national security might be at risk, or the lives of military personnel and other innocent men, women, and children might be in jeopardy. In the case of the game you are about to play, you secretly work for the government of Barunja and are working undercover as a trained spy. It is vitally important that your team members not discover your true mission. Your mission is to PROTECT the scud missiles from being found by deceiving other members of the team about the true location of the missile bunkers. If the missile bunkers are located and destroyed, it could mean the deaths of thousands of innocent people in those areas, as well as escalation of conflict that may result in use of nuclear weapons.*

All participants were seated in front of wireless notebook computers that delivered instructions about the game, collected search and strike results, and administered pre-test and post-test measures. Those in the FtF modality were seated at a round table facing one another. Game decisions were recorded on their computer but all communication among members took place orally. In the audio and text conditions, group members were physically separated and interacted via closed circuit audio or via a text chat window on the computer.

After successful completion of the five-round game of BunkerBuster, each participant completed a post-test survey that asked participants about their experience playing the game, their ratings of communication quality, and perceived task effectiveness. Participants were debriefed, including discussion of the deceptive role, and thanked for their participation.

Dependent Measures The communication quality ratings consisted of the 13 communication quality measures that were reduced through principal components factor analysis to the three meta-dimensions: relational, interaction, and task quality. These three meta-dimensions comprised various dimensions of communication quality that were derived from previous research on CMC (Burgoon et al. 2002a; Stoner 2001). The dimensions were measured by presenting participants a brief definition of the dimension followed by 7-interval unipolar adjective pairs (e.g., "not at all" to "very involved"). Higher scores reflect higher degrees of the quality. Cronbach alpha reliability was computed for the three meta-dimensions.

Relational Quality This communication quality meta-dimension addresses the personal relationships between participants. It includes measures of involvement, connectedness, similarity, openness, positivity, composure, and persuasiveness ($\alpha = .79$).

Interaction Quality This measure gauges the team's ability to coordinate and execute the sharing of information during the task. It includes interaction coordination (how well conversation was coordinated, smooth, and fluent), communication appropriateness, expectedness (communication typicality), and richness of the communication itself ($\alpha = .73$).

Task Quality This meta-dimension measures the effectiveness, efficiency, and task focus of the group's communication. It includes task orientation (percentage of communication related to completing the task), efficiency, and level of critical analysis/feedback ($\alpha = .59$). The marginal reliability on this dimension forecasted weaker and possibly nonsignificant effect sizes for this meta-dimension.

Performance Measures Performance measures for the groups were assessed via actual, objective team performance on the task as well as subjective, perceived performance ratings. Perceived decision effectiveness was measured with items related to how well the individuals felt the group completed the task and how well the group worked together ($\alpha = .70$).

Actual team performance was based on each team's game score, which computed the total number of correct target "hits" by the team members divided by the total number of judgments made. If each team member selected 3 targets and each had 2 targets correct, the final score for a four-person team would be 8/12, or 67. Game scores could range from 0 (all incorrect) to 1.0 (all correct).

BunkerBuster Results

Modality and Communication Quality H1a posited ordinal increases in relational and interactional qualities from text to FtF to audio. The hypothesis was tested on the mean group scores on the meta-dimensions. Planned Helmert contrasts using weights of −2 for text, +1 for FtF and +1 for audio for the first contrast produced significant effects for relational qualities ($t_{(47)} = 1.88$, $p < .05$ one-tailed, $\eta^2 = .13$) and interactional qualities ($t_{(47)} = 2.32$, $p < .05$, $\eta^2 = .10$). The second contrast using weights of −1 for audio and +1 for FtF found a significant effect only for the relational dimension, $t_{(30)} = -2.16$, $p = .039$, $\eta^2 = .13$. Audio and FtF together produced higher ratings than text on both dimensions. Audio produced a higher mean rating than FtF on the relational dimension. Hypothesis 1b posited an ordinal increase for task qualities. The contrast was not significant. Means are shown in Table 9.2.

H2 posited that deception differs from truth on relational, interactional, and task qualities. To prevent deceivers' ratings from influencing group means, group means for the three meta-dimensions were calculated with the space role ratings omitted. Results from the 2 (deception) × 3 (modality) factorial MANOVA analysis failed to find significant main effects for deception on the three meta-dimensions, $F_{(3,}$

$_{42)} = .38, p = .77$, partial $\eta^2 = .03$. See Table 9.3 for means and standard deviations. H2 failed to receive support. Deception did not affect the perceptions of relational, interaction, or task quality. Put differently, any differences in deceivers' communication went undetected or did not adversely affect the group's interaction patterns in a way that affected their communication quality ratings.

Communication Quality and Task Performance Task performance was assessed via perceived task performance and objective task performance.

Perceived task performance H3 posited that greater communication quality meta-dimensions are positively associated with perceived task performance. The hypothesis received full support. The Pearson product-moment correlation between task qualities and perceived performance was positive, $r_{(140)} = .56, p < .01$, two-tailed. Team members' perceptions of the task-related aspects of their communication were positively related to satisfaction with task performance. The other two dimensions were also significant (relational quality $r_{(140)} = .61, p < .01$, two-tailed; interactional quality $r_{(140)} = .46, p < .01$, two-tailed), showing a similar positive and significant association between higher-quality communication and perceived task performance.

A linear regression was performed to evaluate the predictive ability of the communication quality ratings on perceived task performance. Task communication quality significantly predicted perceived performance, $b = .64, t_{(46)} = 3.16, p < .05$. The relational communication quality did not reach conventional levels of significance but was in the correct direction, $b = .37, t_{(46)} = 1.81, p = .08$. The interactional dimension was not significant. The three communication quality meta-dimensions also explained a significant proportion of the variance in perceived performance measures, $R^2 = .47, F_{(3, 46)} = 13.53, p < .01$.

Objective task performance H4 stated that in truthful conditions, communication quality is positively correlated with actual decision quality, whereas in deceptive

Table 9.2 Means and standard deviations for communication quality by modality

Quality	Modality	Mean	Std. Deviation
Relational	Text	5.25	.74
	Audio	5.74	.50
	FTF	5.40	.40
	Total	5.47	.60
Interaction	Text	5.18	.73
	Audio	5.62	.52
	FTF	5.60	.58
	Total	5.46	.64
Task	Text	5.40	.97
	Audio	5.85	.63
	FTF	5.66	.62
	Total	5.64	.78

Table 9.3 Bunkerbuster communication quality means and standard deviations by condition

Dimension	Condition	N	Mean	Std. dev.
Relational	Truth	35	5.49	.57
	Deception	15	5.43	.70
Interaction	Truth	35	5.50	.62
	Deception	15	5.36	.72
Task	Truth	35	5.66	.77
	Deception	15	5.57	.82

Table 9.7 Mean game score by experiment and by condition

Study	Condition	N	Mean	Std. dev.
BunkerBuster	Truth	35	.77	.22
	Deception	15	.36	.38
StrikeCOM	Truth	46	.38	.24
	Deception	47	.20	.16

conditions, the meta-dimensions are negatively associated with decision quality. To place these results in context, we first examined the impact of modality and deception on actual task performance. A 2×3 factorial ANOVA on game scores produced a significant main effect for deception, $F_{(1, 44)} = 21.80$, $p < .01$, $\eta^2 = .32$. Games with deceivers present had worse game scores than groups where deception was absent; see Table 9.7 for the means and standard deviations. There was no main effect for modality or deception by modality interaction. Thus, the presence of deception clearly undermined performance.

To test the hypothesized relationship of communication quality with performance, bivariate correlations were performed separately within the truth and deception conditions. They revealed that in the truthful condition, none of the communication quality dimensions were significantly associated with objective performance, but in the deceptive condition they were. As shown in Table 9.4, the more the group achieved high-quality communication in the deceptive condition, the better they performed. Conversely, poorer communication qualities were associated with poorer performance on the game. Thus, where the presence of deception adversely affected the communication process, it also undermined team performance. Where teams were able to establish effective communication by achieving high involvement and mutuality, by maintaining a smooth and efficient interaction, and by fulfilling task-related responsibilities despite deceit, they were able to mitigate the influence of deception.

Scatterplots were used to further examine the game scores against each of the three communication quality dimensions. The plots revealed, first, that three of the teams in the deceptive condition attained perfect game scores. This could not have happened if the deceivers had chosen the three alternative strike locations that they were told to advocate. These deceivers either did not understand their instructions or abandoned them at the point of the final strike plan. Because these same groups also rated their communication favorably, they contributed strongly to the positive correlation between communication qualities and performance. Removal of these three groups attenuated the relationships but they remained positive.

The scatterplots also revealed that most of the truthful groups not only had moderate to high game scores but also moderate to high communication quality ratings. The net result was a restriction in range that prevented statistical relationships from emerging.

Correlations computed with the combined data set also produced a significant relationship between task communication qualities and actual performance, $r_{(48)} = .33, p < .01$, one-tailed. This result indicates that task-related communication as perceived by team members was an accurate gauge of their team's actual performance.

Discussion of BunkerBuster Results

Analysis of the BunkerBuster data indicated several interesting findings. As predicted in Hypothesis 1, the audio modality produced the highest ratings for relational communication qualities, followed by FtF and lastly synchronous text. For interaction communication qualities, audio and FtF outperformed text. The task communication qualities dimension followed a similar, ordinal pattern. Groups in the audio channel were able to advance relational and interaction-related communication at least as well as groups in the FtF and the text modalities.

Contrary to Hypothesis 2, the presence of deception did not have a significant effect on the communication qualities of the groups. On average, there was no significant difference between the communication qualities between deceptive and non-deceptive groups. However, teams in the deceptive condition scored significantly lower on the task than did teams in the non-deceptive condition. These combined results indicate that deceivers managed to exert influence on the group's decisions without such influence being registered in the team's communication patterns. Put differently, deceivers' communication did not give them away. This means that individuals with ulterior motives can sabotage group work without necessarily being noticed through the quality of communication.

That said, communication qualities were linked to perceived and actual team performance. Consistent with Hypothesis 3, groups with better relational,

Table 9.4 Pearson product-moment correlations between communication quality dimensions and task performance

	Game score, truth Condition ($N = 35$)	Game score, deception Condition ($N = 15$)
Relational quality	−.008	.468*
Interaction quality	−.041	.519*
Task quality	.202	.596**

*correlation significant at the 0.05 level, one-tailed; **correlation significant at the 0.01 level, one-tailed

interactional and task communication perceived their team's decision-making as effective. Communication quality also affected actual performance within groups where deception was present. Those groups whose relational, interactional, and task communication quality was lower also performed worse, whereas those groups who succeeded in expressing such relational messages as receptivity, connectedness, positivity, and involvement; whose interactions were more coordinated, expected, appropriate, and composed; and whose task-related messages were efficient yet included ample analysis and evaluation, were able to achieve better decisions, despite the presence of a member working against the common goal.

This pattern was contrary to Hypothesis 4 in that we posited a positive relationship between communication qualities and objective performance in the truthful conditions and a negative relationship in the deception conditions. The lack of significant correlations in the truthful conditions was most likely due to a restricted range on the measures, which prevents relationships from being discovered. However, in the deception condition, in retrospect, it is understandable that groups that succeeded in overcoming negative influences within their group would have performed better and those that were "dragged down" by their dissembling member would have done worse. Had only the communication of the deceiver been analyzed, our hypothesis might have held, but the communication qualities were examined as a group, allowing the nondeceptive team members' communication and perceptions to drive the results.

Because not all deceivers appeared to follow instructions, and because the group sizes were unbalanced due to failure of some participants to show up and complete the four-person groups, the next experiment sought to improve upon instructions to deceivers and to increase sample size. The sample size issue was addressed by reducing group size to three-person teams. Additionally, a large number of improvements were made in the game, which morphed into a new and flexible networked, multiplayer computer game designed and built by the Center for the Management of Information (CMI) at the University of Arizona (Twitchell et al. 2005).

StrikeCOM Methods

Participants Participants ($N = 285$) were undergraduate students recruited from an introductory Management Information Systems class required for all business students. Participants received extra credit in the course in exchange for participating.

Procedures and Independent Variables The design was a 2 (deceptive versus non-deceptive) × 3 (audio versus text-based versus FtF modality) mixed model factorial experiment.

Participants played StrikeCOM, a game that was patterned after BunkerBuster and required teams to find targets or information hidden on a game board within a predefined number of turns. Participants were randomly assigned to roles and conditions. The three roles commanded space, air, or human intelligence assets. The deception manipulation was modeled after the one used in BunkerBuster. In this version, the person in the space role again received the deception induction.

The same three modalities were tested. Those in the synchronous text condition were situated in separate rooms and communicated through text chat that was recorded to the StrikeCOM database. Those in the audio modality were likewise separated and used headphones and microphones to communicate over the closed-circuit network. Those in the FtF condition were seated around a round table, facing each other. Interactions of these groups were recorded using digital camcorders.

The participants completed a pre-test survey to assess demographics and computer and group work experience. All participants then viewed an instructional video, delivered by laptop. Participants in the deceptive condition received their special instructions at the end of their video. The teams played the game to completion and completed a post-test survey. The post-test survey included the same measures as the BunkerBuster post-test survey. The subjects were debriefed, including discussion of the role of deception, and thanked for their participation.

Dependent Measures The same communication quality meta-dimensions, perceived effectiveness, and game scores that were utilized in BunkerBuster were utilized in the StrikeCOM experiment.

StrikeCOM Results

Modality and Communication Quality Analysis of H1a and H1b, which posited that relational, interaction, and task communication quality would decrease across modalities from audio and FtF to text, produced a multivariate main effect for modality, $F_{(6, 174)} = 6.81, p < .01$, partial $\eta^2 = .19$. Each of the communication quality dimensions was significant, relational quality, $F_{(2, 89)} = 4.83, p < .01, \eta^2 = .10$, interaction quality, $F_{(2, 89)} = 12.42, p < .01, \eta^2 = .22$, and task quality, $F_{(2, 89)} = 18.00$, $p < .01, \eta^2 = .29$. This omnibus test indicated a strong effect of modality on the combined communication qualities. The means shown in Table 9.5 revealed that the FtF modality ranked higher than audio on two of the dimensions. Next, the comparison of the text condition to the combined FtF and audio conditions was decomposed into Helmert contrasts with 2 degrees of freedom. The first contrast compared text (−2) to audio (+1) and FtF (+1). The second contrast compared audio (−1) to FtF (+1). The results indicated a significant difference between text and the other two modalities for relational qualities, $t_{(92)} = 2.91, p < .01, \eta = .08$; interaction qualities, $t_{(92)} = 4.88, p < .01, \eta = .21$; and task qualities, $t_{(92)} = 5.97, p < .01, \eta = .28$. The hypothesis was supported. An ordinal increase from text to FtF to audio did not materialize in this case.

The second contrast was not significant for any of the dimensions (relational $p = .10$, interaction $p = .07$, task $p > .10$).

Deception and Communication Quality H2, that deception and truth differ on the relational, interaction, and task qualities, was tested in the same 2 (deception) × 3 (modality) factorial MANOVA, with the three communication quality dimensions as the dependent measures. No significant main effects were found between the deception and control conditions on relational, interaction, or task quality meta-dimensions, $F_{(3, 87)} = .59$, $p = .62$, partial $\eta^2 = .02$. H2 failed to receive support. The means and standard deviations for the communication quality dimensions by condition are found in Table 9.6. Like BunkerBuster, deception did not lower perceptions of relational, interaction, or task quality.

Communication Quality and Task Outcomes Similar to BunkerBuster, task performance was assessed via perceived task performance and objective task performance.

Perceived task performance H3 posited that greater communication quality meta-dimensions are associated with more favorable perceived task performance. The bivariate correlations conducted on individual ratings were positive and significant, relational quality $r_{(140)} = .56$, $p < .01$, two-tailed; interactional quality $r_{(140)} = .54$, $p < .01$, two-tailed; task quality $r_{(140)} = .53$, $p < .01$, two-tailed. H3 was supported; positive relational, interaction, and task communication was positively associated with perceived task performance.

Objective task performance H4 stated that in truthful conditions, communication quality is positively correlated with decision quality whereas in deceptive conditions, the meta-dimensions are negatively associated with decision quality. A 2 × 3 factorial ANOVA on game scores produced a significant main effect for deception, $F_{(1, 92)} = 17.42$, $p < .01$, $\eta^2 = .16$. Games with deceivers present had worse game scores than groups where deception was absent; see Table 9.7 for the game score means and standard deviations.

Bivariate correlations were performed within the truthful and deceptive conditions. Under truth, the relational quality dimension was positively correlated with objective task performance and the interaction quality and task quality dimensions

Table 9.5 Means and standard deviations for communication quality by modality

Quality	Mode	Mean	Std. dev.
Interaction quality	Text	5.05	0.66
	Audio	5.54	0.51
	FTF	5.76	0.55
Relational quality	Text	5.33	0.61
	Audio	5.59	0.51
	FTF	5.78	0.55
Task quality	Text	5.14	0.65
	Audio	5.85	0.47
	FTF	5.83	0.49

showed similar trends, consistent with H4. But contrary to the hypothesis, under deception, the interaction and task quality dimensions also produced positive relationships: under truth, relational quality and performance, $r_{(44)} = .32, p < .05$, one-tailed; under deception, interaction quality and performance, $r_{(45)} = .29, p < .05$, one-tailed; task quality and performance, $r_{(45)} = .32, p < .05$, one-tailed. Correlations computed with the combined data set also produced a significant relationship between task communication qualities and actual performance, $r_{(93)} = .27, p < .05$, two-tailed.

This result indicates that task-related communication as perceived by team members was an accurate gauge of their team's actual performance. Likewise, a significant relationship emerged between interaction coordination and actual performance, $r_{(93)} = .23, p < .05$, two-tailed, indicating that a team's perception of the ability to interact well during the task was also an accurate reflection of the team's ability to perform on the task. Correlations computed using a dataset split by deception identified a significant relationship between subjective performance and objective performance in the truthful condition, $r_{(46)} = .56, p < .01$, one-tailed. No significant correlation was found in the deceptive condition, $r_{(47)} = .12, p = .20$.

Discussion of StrikeCOM Results

The first hypothesis investigated the effects of modality interactivity on the communication quality dimensions. H1 predicted that text would receive the lowest ratings on the communication quality dimensions, followed by FtF and audio. This hypothesis was supported. There was no significant difference between the audio-only and FtF ratings. The presence of nonverbal cues afforded by proximity in the FtF condition did not significantly improve or impair perceived quality of team communication.

H2 posited that deception would alter the communication quality of the team as compared to teams with no deceivers present. This hypothesis received no support as there was no significant difference between truthful and deceptive conditions. Deceivers were able to perpetrate the deception without any noticeable difference in the perceived communication quality of the team. Communication qualities, however, were linked to both the subjective and objective performance of the teams.

Table 9.6 Strikecom communication quality means and standard deviations by condition

Dimension	Condition	N	Mean	Std. dev.
Interaction	Truth	47	5.47	.66
	Deception	48	5.40	.64
Relational	Truth	47	5.57	.59
	Deception	48	5.55	.59
Task	Truth	47	5.65	.65
	Deception	48	5.54	.63

H3 posited that perceptions of group performance would be positively associated with the communication quality dimensions. As predicted, this hypothesis received full support. All three communication quality dimensions were positive and significantly correlated with *perceived* performance on the experimental task.

H4 stated that under truthful conditions, the communication dimensions would be positively associated with objective score. Alternatively, when deception was present, the communication quality would be negatively related to the objective score of the team. The results indicated that in the truthful condition, only the relational quality dimension was positively associated with objective performance. Teams that were able to produce higher levels of team cohesiveness performed better on the experimental task. Under deception, interactional and task qualities were both positive and significantly associated with objective performance. These findings are contrary to the hypothesized relationship. Groups that were able to overcome the deceivers' subterfuge and communicate well as a team were able to achieve higher scores. Teams that were not able to overcome the burden of the deception on team communication did not perform as well on the task.

The analysis of team members' perceptions of team performance versus objective performance also shows the impact of deception. In truthful conditions, a positive, significant correlation existed between subjective performance assessments and objective performance. However, in the deceptive condition, no significant correlation emerged. These results further indicate that deceivers are able to sabotage the team efforts, resulting in conflicting self-assessment reports regarding team performance. The deceptive teams scored lower on the objective measure and the members of the team were not able to accurately assess performance.

General Discussion

The current research sought to replicate and expand upon previous findings regarding the relationship of deception to communication quality and group task outcomes. These issues are increasingly salient as much more daily conversation is transacted via mediated formats. The principle of interactivity and the communication quality model present a framework for analyzing and understanding communication modalities and other external, contextual factors on group communication processes and task outcomes. The communication quality model provides an explanatory mechanism to measure and examine the mediating effect of communication qualities on team outcomes such as decision quality, task efficiency, and social judgments.

The first hypothesis examined the effects of modality interactivity on the three communication quality dimensions. Of the three modalities examined, synchronous text is the least interactive and FtF communication is the most interactive. Contrary to the usual prediction that FtF is preferred for high-quality communication and task outcomes, we predicted that the audio modality would equal or exceed FtF interaction in generating high-quality task communication and that text communication would produce the lowest quality. Consistent with previous research (Burgoon et al. 2003), the audio modality either exceeded or matched FtF communication in terms of relational, interaction, and task qualities in both the BunkerBuster and StrikeCom data. The implication to be drawn is that loss of proximity that accompanies the audio condition does not impair the ability to build involvement or cohesiveness on the team. Neither does the lack of proximity impair smooth, appropriate, and tension-free interaction or the efficient and effective exchange of information, analysis, and evaluation. The audio condition is sufficient for enabling the smooth coordination of information exchange. Of course, these findings are context dependent. The results depend on the composition of the team, the size of the team, and the nature of the task. For example, with much larger group sizes, such as large team videoconferencing or distance education with large classes, the situation would likely change because the coordination of turn-taking and distinguishing different speakers becomes more challenging. Additional research is needed to further investigate the impact of task type on modality interactivity and communication quality ratings.

As expected, the text-based modality received the lowest ratings for both relational and interaction quality. These results are consistent with the principle of interactivity in that the lack of multiple modes to send cues, the lack of proximity, and the decreased level of synchronicity impaired the ability of the team to build cohesiveness, connection, and positive interpersonal relationships as well as to coordinate smooth message exchange.

The presence of deception in a group was predicted to alter the relational and interaction factors as compared to the control condition. Previous research had produced mixed results regarding this relationship. The present research found no significant difference on average between the deceptive and nondeceptive teams on the communication quality ratings. These results can be seen as a positive or as problematic. Since deception in some shape or form is so prevalent in everyday discourse, it is encouraging that the presence of deception need not impair the quality of the group's communication. Groups can still foster involvement, mutuality, similarity, and coordinated message exchange despite the presence of deception. However, from a diagnostic standpoint, communication qualities did not prove to be a marker of deceptive communication; deceivers were successful at perpetrating their deception without naïve team members recognizing their ulterior motives through their communication. Deceivers may even capitalize on the group's communication patterns to achieve their own ends. In groups that are struggling to collaborate and communicate, the deceiver may commit the deceptive act by providing sparse details that are difficult to understand. However, in groups that are achieving high levels of communication quality, the deceiver may use the opposite approach and provide ample information to present a credible appearance and blend in with the team's existing communication norms. The level of interactivity afforded by the modality impacts the ability of the deceiver to execute these approaches.

The next set of hypotheses examined the relationship between the communication quality meta-dimensions and interaction outcomes. The third hypothesis predicted that relational and interaction coordination quality is positively associated with perceived task effectiveness. Research has shown that relational dimensions are positively associated with social judgments of team members. An extension of the correlation with positive social judgments is an overall positive assessment of group performance. Positive assessments of the group's communication processes were positively correlated with perceived task performance. These results suggest a possible halo effect whereby participants who view their team as having robust communication quality also feel their team performed well on the task.

The last hypothesis examined the relationship between communication qualities and the group's objective performance. Previous research had indicated partial support for the notion that relational qualities are positively correlated with objective performance under truthful conditions and negatively associated with performance under deceptive conditions. The current findings do not coincide with this previous research. Interestingly, it is the deceptive condition that is positively correlated with objective performance. The deceivers were able to successfully undercut the teams, as there were significant differences between truthful and deceptive game scores. In both experiments, the teams that had no deceptive member (i.e., the control condition) significantly outperformed the deceptive condition teams. The correlation indicates that some of the groups in the deceptive condition were able to overcome the adverse effects of the deceivers and were able to perform better in the experimental task. In groups that had lower communication quality dimensions, lower objective scores were reported. Further research is needed to investigate these relationships, utilizing a different objective performance metric, inasmuch as the current ones may have suffered from a truncated range.

Additional work is needed to further understand the role of interactivity in the communication process, especially when deception is present. Here, naïve group members failed to detect deception when present. It is plausible that the communication quality and performance would change as groups re-convened to complete additional tasks. Investigating the relationship of trust and suspicion to the three meta-dimensions would further inform how the communication patterns relate to deception detection and team performance.

Acknowledgement We are grateful to the Army Research Office for funding much of the work reported in this book under Grant W911NF-16-1-0342.

Funding Disclosure This research was sponsored by the Army Research Office and was accomplished under Grant Number W911NF-16-1-0342. The views and conclusions contained in this document are those of the authors and should not be interpreted as representing the official policies, either expressed or implied, of the Army Research Office or the U.S. Government. The U.S. Government is authorized to reproduce and distribute reprints for Government purposes notwithstanding any copyright notation herein.

References

Biocca, F., Harms, C., & Burgoon, J. K. (2003). Toward a more robust theory and measure of social presence: Review and suggested criteria. *Presence: Teleoperators & Virtual Environments, 12,* 456–480.

Buller, D., & Burgoon, J. K. (1996). Interpersonal deception theory. *Communication Theory, 6,* 203–242.

Buller, D. B., Burgoon, J. K., Buslig, A., & Roiger, J. (1996). Testing interpersonal deception theory: The language of interpersonal deception. *Communication Theory, 6,* 268–289.

Burgoon, J. K., Bonito, J. A., Bengtsson, B., Ramirez, A., Dunbar, N., & Miczo, N. (2000). Testing the interactivity model: Communication processes, partner assessments, and the quality of collaborative work. *Journal of Management and Information Systems, 16*(3), 33–56.

Burgoon, J. K., Bonito, J. A., Ramirez, A., Jr., Dunbar, N. E., Kam, K., & Fischer, J. (2002a). Testing the interactivity principle: Effects of mediation, propinquity, and verbal and nonverbal modalities in interpersonal interaction. *Journal of Communication, 52,* 657–677.

Burgoon, J. K., Burgoon, M., Broneck, K., Alvaro, E., & Nunamaker, J. F. (2002b). *Effects of synchronicity and proximity on group communication.* Paper presented at the annual convention of the National Communication Association, New Orleans.

Burgoon, J. K., Stoner, G. M., Bonito, J. A., & Dunbar, N. E. (2003, January 06–09). *Trust and deception in mediated communication.* Paper presented at the 36th annual Hawaii international conference on system sciences, Big Island, Hawaii.

Burgoon, J. K., Blair, J. P., & Strom, R. (2008). Cognitive biases, modalities and deception detection. *Human Communication Research, 34,* 572–599.

Burgoon, J. K., Chen, F., & Twitchell, D. (2010). Deception and its detection under synchronous and asynchronous computer-mediated communication. *Group Decision and Negotiation, 19,* 346–366.

Carlson, J., & Zmud, R. (1999). Channel expansion theory and the experiential nature of media richness perceptions. *Academy of Management Journal, 42,* 153–170.

Daft, R. L., & Lengel, R. H. (1986). Organizational information requirements, media richness and structural design. *Management Science, 32,* 554–571.

Dennis, A. R., & Kinney, S. T. (1998). Testing media richness theory in the new media: The effects of cues, feedback, and task equivocality. *Information Systems Research, 9,* 256–274.

Dennis, A. R., & Valacich, J. S. (1999). *Rethinking media richness: Towards a theory of media synchronicity.* Paper presented at the proceedings of the thirty-second annual Hawaii international conference on system sciences.

Dennis, A. R., Valacich, J. S., Speier, C., & Morris, M. G. (1998). *Beyond media richness: An empirical test of media synchronicity theory.* Paper presented at the proceedings of the thirty-first annual Hawaii international conference on system sciences.

Dennis, A. R., Fuller, R. M., & Valacich, J. S. (2008). Media, tasks, and communication processes: A theory of media synchronicity. *MIS Quarterly, 32,* 575–600.

Ekman, P. (1996). Why don't we catch liars? *Social Research, 63,* 801–818.

George, J. F., Giordano, G., & Tilley, P. (2016). Website credibility and deceiver credibility: Expanding Prominence-Interpretation Theory. *Computers in Human Behavior, 54,* 83–93.

Hiltz, S. R. (1988). Productivity enhancement from computer-mediated communication: A systems contingency approach. *Communications of the ACM, 31,* 1438–1454.

Nardi, B. A. (2005). Beyond bandwidth: Dimensions of connection in interpersonal communication. *Computer Supported Cooperative Work, 14,* 91–130.

Ramirez, A., Jr., & Burgoon, J. K. (2004). The effect of interactivity on initial interactions: The influence of information valence and modality and information richness on computer-mediated interaction. *Communication Monographs, 71,* 422–447.

Rice, R. E. (1992). Task analyzability, use of new media, and effectiveness: A multi-site exploration of media richness. *Organization Science, 3,* 475–500.

Serota, K. B., Levine, T. R., & Boster, F. J. (2010). The prevalence of lying in America: Three studies of self-reported lies. *Human Communication Research, 36*, 2–25.

Stoner, G. M. (2001). *Decision-making via mediated communication: Effects of mediation and time pressure* (Unpublished masters thesis). University of Arizona, Tucson, AZ.

Twitchell, D. P., Wiers, K., Adkins, M., Burgoon, J. K., & Nunamaker, J. F., Jr. (2005). *Strikecom: A multi-player online strategy game for researching and teaching group dynamics.* Paper presented at the proceedings of the 38th annual Hawaii international conference on system sciences.

Vickery, S. K., Droge, C., Stank, T. P., Goldsby, T. J., & Markland, R. E. (2004). The performance implications of media richness in a business-to-business service environment: Direct versus indirect effects. *Management Science, 50*(8), 1106–1119.

Walther, J. B. (1992). Interpersonal effects in computer-mediated interaction: A relational perspective. *Communication Research, 19*, 52–90.

Walther, J. B. (1996). Computer-mediated communication: Impersonal, interpersonal, and hyperpersonal interaction. *Communication Research, 23*, 1–43.

Chapter 10
Incremental Information Disclosure in Qualitative Financial Reporting: Differences Between Fraudulent and Non-fraudulent Companies

Lee Spitzley

Introduction

Financial statement fraud is the act of intentionally misstating the true condition of a firm's financial health. Firms may use misrepresentations, concealments, or non-disclosures to achieve some material benefit, either for the company or to enrich the individuals personally (Dyck et al. 2013). This serious infraction damages the firms involved, investor trust in markets, and is often difficult to identify. Financial statement fraud comprises about 5% of all accounting fraud cases, but they are the most costly (Association of Certified Fraud Examiners 2014). Dyck et al. (2013) estimated that between 5.6% and 14.5% of firms are engaging in fraud at any given time, with a cost of 20.4% of the enterprise value of these firms. For frauds that occurred between 1978 and 2002, legal fines averaged $23.5 million, and market value penalties were over 7.5 times higher than legal fines (Karpoff et al. 2008). Although fraud occurs frequently, it is difficult to identify and requires significant resources to investigate.

Researchers and practitioners have found that qualitative disclosures—discussions that provide a narrative to accompany financial results—have characteristics useful for fraud detection. In the financial reporting environment, several valuable sources of narrative disclosure provide qualitative interpretation of financial data. Two frequently studied venues are the earnings conference call and the Management's Discussion and Analysis section of a 10-K or 10-Q report (MD&A). At the end of each quarter, most companies host an earnings conference call, where a small panel of executives discuss the preceding quarter with investment analysts who follow the company and provide recommendations. These calls are a major information event, and they contain meaningful information beyond the accompanying press release

L. Spitzley (✉)
University at Albany, SUNY, Albany, NY, USA
e-mail: lspitzley@albany.edu

© Springer Nature Switzerland AG 2021
V. S. Subrahmanian et al. (eds.), *Detecting Trust and Deception in Group Interaction*, Terrorism, Security, and Computation,
https://doi.org/10.1007/978-3-030-54383-9_10

(Matsumoto et al. 2011). They normally begin with prepared remarks from the executives, which are usually written in conjunction with the firm's legal team. A question-and-answer session with investment analysts who report on the company follows the prepared portion. Analysts will ask questions about areas they believe to be most important or are unclear. During these calls, fraudulent executives tend to use more extreme positive language and discuss shareholder value less frequently (Larcker and Zakolyukina 2012), and also have a higher tendency to script responses to analyst questions (Lee 2016).

Another source of qualitative information about firm performance is the Management's Discussion and Analysis (MD&A) section of quarterly and annual financial statements submitted to the SEC. The MD&A is an SEC-required section of the financial statements, and it provides a narrative discussion of the financial results of the period. It contains qualitative and forward-looking statements about the business, supplemented by tables and figures. These statements are carefully prepared and vetted by a legal team, yet differences in the linguistic cues of fraudulent companies still remain (Humpherys et al. 2011). Fraudulent companies also spend less time discussing information related to governance, financial constraints, and revenues (Hoberg and Lewis 2014).

Studies that use narrative disclosures to identify fraud tend to focus on a single reporting venue (Hoberg and Lewis 2014; Humpherys et al. 2011; Larcker and Zakolyukina 2012). One limitation of this approach is that it does not consider the relationships between the multiple disclosures that cover a single fiscal period. Firms often release these disclosures at different times, leaving managers the opportunity to adjust their story between disclosures (Davis and Tama-Sweet 2012). Whether or not fraudulent firms modify their stories differently than non-fraudulent firms remains unknown. Dissimilarity between narratives is an indicator of greater information disclosure (Brown and Tucker 2011; Davis and Tama-Sweet 2012; Lee 2016), which may increase attention from investors and regulators. Fraudulent managers may use greater similarity between narratives to reduce the risk of presenting contradictory information that comes from maintaining a complex deception, particularly during the Q&A of the calls. To reduce inconsistencies and create a favorable impression, it is possible that fraudulent companies will disclose less information across reporting venues. Repetition can lead to a more persuasive message (Cacioppo and Petty 1989; Petty and Cacioppo 1979). Language similarity may also indicate that managers take a more active role in the document preparation process. This would reduce the number of people who interact with the area of fraud. This leads to the overall research question:

Relative to non-fraudulent companies, do the CEOs and CFOs of fraudulent companies use language that is more similar between their conference calls and MD&A sections than the CEOs and CFOs of non-fraudulent companies?

To answer this question, I present a novel approach to identifying fraudulent financial statements by considering content modifications in narratives covering a single reporting period (i.e. fiscal quarter). This study measures language similarity between the earnings calls and subsequent MD&A section. Earnings calls present an interesting opportunity to investigate prepared statements and spontaneous remarks from the question and answer session with analysts. Figure 10.1 contextualizes this research and similar studies using document comparison in financial analysis.

Answering this question has implications for both the deception and accounting domains. In deception research, it is often difficult to create high-stakes deception in an experimental setting and obtaining real-world data with known cases of truth and deception is challenging[1]. Also of interest to deception researchers are story-consistency strategies by deceptive parties when there is ample time to prepare.

From an accounting and fraud investigation perspective, this research expands on knowledge of strategic corporate reporting. Analyzing multiple disclosures provides a chance to examine variations between truth-tellers and deceivers at multiple times. This will also improve understanding of strategic corporate financial reporting by learning how fraudulent and non-fraudulent firms differ in their information disclosure strategies between venues.

From a practical perspective, there is a growing interest in the analysis of qualitative information to identify fraud in public companies. This non-financial information provides a complementary source of information to the quantitative analysis of financial measures. For example, the SEC is increasing its focus on the MD&A section of annual reports because of their ability to distinguish fraudulent activities (Association of Certified Fraud Examiners 2013).

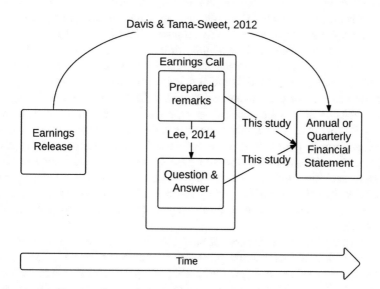

Fig. 10.1 Studies comparing multiple disclosure venues

---

[1] I use Accounting & Auditing Enforcement Releases (AAERs) to establish ground truth.

The remaining portion of this paper is as follows: In Section "Research Development", I summarize the relevant background literature on financial fraud and prior methods used to identify fraudulent companies. Section "Method" describes the research hypotheses. In section "Analysis", I describe the data collection process and my resulting sample. Following that is an analysis of the data (section "Discussion"), and I conclude with a discussion of the results and avenues for future study.

Research Development

Disclosure Strategy and Impression Management

Impression management is the "process by which people control the impressions others form of them" (Leary and Kowalski 1990, p. 34). There are two major components of this process: impression construction and impression motivation. Impression construction is the type of impression one is trying to create, and impression motivation is the level of how motivated someone is to control how others see them. While this original description was defined in terms of individuals, this definition of impression management is also useful in understanding how companies strategically present information through their disclosures (Lo and Rogo 2014; Merkl-Davies et al. 2011; Merkl-Davies and Brennan 2007).

Narrative disclosures present an opportunity for impression management as a way to advance their goals (Merkl-Davies and Brennan 2007), and firms do use these opportunities to present a positive image (Merkl-Davies et al. 2011). When there is bad news to report, for example, firms have a tendency to withhold this information from a relatively transparent reporting venue (earnings press release) and place it in the text of the financial statement (Davis and Tama-Sweet 2012). Companies that experience a change in CEOs will strategically use the presentational graphs in financial statements to create a favorable image of their performance (Godfrey et al. 2003). In the earnings calls with analysts, managers in firms that are performing poorly or are at risk of lawsuits tend to script their responses to analysts during the question and answer portion of the call, presumably to mitigate the risk of accidental information disclosure (Lee 2016). Another way to portray a desired identity is by maintaining consistency; the language in financial statements seems to be consistent with a firm's reported financial state (Merkl-Davies et al. 2011).

In the case of fraud, managers not only need to use impression management to portray their desired image—they must also portray a persuasive image that leads their audience to believe that what they are reporting reflects the true results of their performance. This additional complexity in impression construction should lead to differences in the ways that fraudulent and non-fraudulent companies present information in narrative disclosures.

Differences Between Groups

When executives are trying to deceive in this context, I assume that they are motivated to convince investors and analysts that their financial information is accurate through impression management techniques. While both fraudulent and non-fraudulent firms engage in impression management, deceptive parties should be more motivated to create a consistent story. Prior research on deception has revealed that unrehearsed liars tend to have more inconsistencies in their language (Walczyk et al. 2009), and these inconsistencies may increase suspicion by other parties involved in the interaction (Buller and Burgoon 1996).

There are several reasons to suspect that fraudulent firms will have language that is more similar between disclosure venues. In this setting, there are two options for persuading the audience of their integrity: discussing only the accurate items (and ignoring the fabricated items, deception by limiting disclosure) or by disclosing fabricated items (deception by fabrication) (Hoberg and Lewis 2014). In the first instance, there should be an increase in similarity because of the limited pool of items to discuss. In the second case, deception by fabrication, there should be increased similarity because of attempts to maintain consistency in the fabricated story.

Deceivers tend to use less-diverse language in spontaneous communication (Zhou and Zhang 2008). However, the use of less-diverse language also appears in the thoroughly-prepared MD&As of fraudulent companies (Humpherys et al. 2011). Less-diverse language may reflect a strategy to keep a manufactured story consistent. Although theories of cognitive load are not likely to apply to the prepared portion of the earnings call, there may be an incentive to keep this portion simple to avoid increasing the difficulty of answering questions during the Q&A. To mitigate the potential of disclosing information that may later be used against the company, at-risk companies tend to use higher amounts of scripting during the Q&A portion of the call (Lee 2016). If this is the case, then scripting should lead to higher similarity between Q&A and MD&A. Another possibility is that managers convey consistency across narratives by incorporating off-script remarks into the MD&A.

The strategies of either limiting information or disclosing fabricated information lead to the following research question:

RQ1: Do fraudulent managers have more similar language than non-fraudulent companies between the conference call and the MD&A (a) for the prepared portion, and (b) for the Q&A portion?

Differences Between CEOs and CFOs

Larcker and Zakolyukina (2012) separate the earnings call speech data by CEO and CFO. Their results show greater predictive ability from CFOs than CEOs. I also separate my data to compare findings from CEOs and CFOs.

Predictive Ability of the Prepared Portion Versus the Q&A

Larcker and Zakolyukina (2012) revealed a counter-intuitive finding: that the prepared remarks and Q&A remarks in an earnings call display similar linguistic variations between fraudulent and non-fraudulent firms. This is interesting, as one might expect a less-rehearsed scenario to have greater predictive ability due to greater cognitive load. It is also possible that the prepared remarks will have a greater predictive ability, since analysts lead the discussion during the Q&A. Bloomfield (2012) calls for further investigation into this issue; therefore, I propose the following exploratory research question:

RQ2: Are there differences in predictive ability between the prepared remarks and Q&A?

Method

This study investigates quarterly earnings calls, rather than earnings press releases, because they contain prepared and spontaneous remarks and provide information above what is contained in the accompanying press release (Matsumoto et al. 2011). The MD&A is a suitable comparison document because companies often release it sometime after the occurrence of the earnings call, giving managers ample time to manage strategically any new information in the MD&A.

Identification of Fraud

There are varying approaches to determining which companies to label as fraudulent (Karpoff et al. 2012). One approach is to use firms SEC identified as fraudulent through an AAER (Cecchini et al. 2010; Dechow et al. 2011; Humpherys et al. 2011; Purda and Skillicorn 2015). Larcker and Zakolyukina (2012) use a wider approach, using companies that had a disclosure of material weakness, an auditor change, a late filing, or a Form 8K filing. Another potential source of fraud identification is the Stanford Securities Class Action Clearinghouse (SSCAC) dataset, which contains many instances of shareholder lawsuits initiated because of fraud. However, this data also includes many other lawsuits not related to fraud, and after eliminating frivolous lawsuits and those occurring in small firms, the size of the sample drops below 250 observations (Dyck et al. 2010).

I use companies identified as fraudulent by the SEC's AAER database because of its general acceptance of identifying fraud, and the fact that it has very few false positives (i.e., companies labeled as fraud that were not fraudulent). I used the database from Dechow et al. (2011), which contained the reason for the AAER and the relevant time periods. This database was current through September 2013. To increase sample size, I followed their procedures and added AAERs from September 2013 through January 2015. The final sample contains fraudulent statements from 2006 through 2013. The selected companies had at least one quarterly statement affected by the fraud. Like many prior studies, I assumed that companies not identified by an AAER were not fraudulent.

Data

Seeking Alpha (seekingalpha.com; hereafter SA) hosts earnings call transcripts and makes them publicly available. Transcribed earnings calls on SA begin appearing for events that took place around 2006, with many others beginning at some time before 2008. For each firm in the AAER database, I attempted to locate the earnings call transcripts that take place during the period of the fraud. There were a total of 139 fraudulent earnings calls available for this period.

For each call, a web scraper separated the utterances from the call by speaker, position, and company (if an analyst). It also separated the calls by part (prepared remarks or Q&A). I collected financial statements from the SEC EDGAR database. A script separated the MD&A sections from the rest of the statement and removed tables from the text.

Control Companies

Because the transcript database has limited coverage, many companies lack enough pre-fraud data to compare within firms. Therefore, I create a comparison group using companies with similar financial characteristics. Table 10.1 shows the similarity criteria. I used the first two digits of the SIC code to find companies from a similar industry, and then narrowed the results further by matching firms with similar market valuations. All firms were matched using data from the quarter when the fraud starts. In the case of multiple firms meeting the criteria, the control company

Table 10.1 Control company selection criteria

Variable	Description	Matching criteria
SIC	Standard Industry Code	Same two-digit SIC
ln(ASSETS)	Natural log of the firm's assets	±15%

was randomly selected. If the selected firm had no transcripts available, a new control company was selected at random until one was found with available transcripts.

Measuring Language Similarity

To compare the earnings calls to their respective MD&A, I first weight the word counts for each document using the term frequency-inverse document frequency (tf-idf) weighting algorithm and then measure similarity using cosine similarity (Manning et al. 2008, pp. 117–125).

Term frequency represents the number of times a term t appears in each document, represented as $tf_{t,\,d}$, which is a $t \times d$ matrix containing the number of times each term appears for each document d. The idf formula is

$$idf_t = \log \frac{N}{df_t},$$

where idf_t is the weighting of term t, N is the number of documents in the corpus, and df_t is the number of documents containing term t.

The multiplication of $tf \times idf$ yields the weighted matrix for the documents in the corpus. Each row vector in this matrix represents a document, and these vectors can be compared to each other using cosine similarity. Because these vectors only contain non-negative values, the value of the cosine similarity will be between 0 and 1, where 0 represents two completely orthogonal vectors (completely dissimilar) or 1, which represents the same vector (completely the same). One useful characteristic of this approach is that the length of the documents is not a factor. The formula for cosine similarity is

$$\cos \theta = \frac{v_1 \cdot v_2}{v_1 v_2},$$

where θ is the angle between vectors v_1 and v_2, the numerator is the dot product of the two vectors, and the denominator is the product of the vector lengths (i.e. $\| v_1 \|$ represents the length of v_1). This function returns a value between 0 and 1.

Financial literature has used tf-idf and cosine similarity measure to compare MD&A sections for modifications over time (Brown and Tucker 2011). One of the primary benefits of tf-idf is that it gives more weight to rare terms in a corpus and greatly diminishes the importance of function words like "the" or "and". Many words have connotations in an everyday setting, such as "debt" or "gain", but mean little in a financial setting, where these terms are frequent (Loughran and McDonald 2011); tf-idf can mitigate this issue.

Text Preparation

For all earnings call turns-at-talk belonging to managers, I identified whether the speaker was the CEO or CFO. I then merged each manager's language for each part of the call. For example, all CEO remarks during a Q&A session would merge into one text document. The Porter stemming algorithm in the Natural Language Tool Kit (NLTK; www.nltk.org/) reduced words in earnings calls and MD&As to their root form.

Following the text cleaning, I computed the idf weights for each word. For each call portion and the MD&A, I created the term frequency matrix, and then computed the tf-idf matrix. I then used cosine similarity to compare the tf-idf vectors of the prepared remarks and Q&A of each call to the corresponding MD&A section. For each combination, I merged the similarity measures with the same-period financial variables from Compustat.

Document Length Correction

Brown and Tucker (2011) showed that similarity will increase as document length increases. To correct for this, they estimated a polynomial regression to get the expected similarity score for a pair of documents. From the expected similarity score they subtracted the actual similarity score to get the residual value. The residualized value is free of the effects of document length. On this new measure, higher values still have the same interpretation: the higher the value, the greater the similarity (respective to what we would have expected for a document of the given length). I also use this adjustment in my analysis.

Analysis

Descriptive Statistics

Table 10.2 presents the sample descriptive statistics. There were fourteen industry groups represented, with many companies coming from the categories of depository institutions and business services (such as IT firms). There were more observations near the middle of the sampling period, reflecting the limited coverage of Seeking Alpha in the earlier years and the latency between fraud occurring and the filing of an AAER in the later years.

There were 275 transcripts on SA. After eliminating transcripts that did not parse properly, quarters that did not have a matched MD&A, and those without complete financial data, the total usable number of observations was 230. The final sample

Table 10.2 Descriptive statistics

Panel A: Means & standard deviations

	Fraud			Non-fraud		
Item	N	Mean	SD	N	Mean	SD
SIMSCORE						
CEO prepared	106	0.281	0.187	111	0.307	0.168
CEO Q&A	107	0.105	0.093	113	0.104	0.077
CFO prepared	100	0.349	0.193	106	0.314	0.157
CFO Q&A	100	0.088	0.084	106	0.087	0.070
ln(MDA_LEN)	114	9.327	0.987	116	9.433	0.879
ln(ASSETS)	114	8.140	3.405	116	7.857	3.107
ln(MVE)	114	7.326	2.434	116	7.255	2.475
ln(AGE)	114	5.180	0.749	116	5.219	0.646
INST_OWN	108	0.711	0.313	93	0.639	0.226

Panel B: Companies by industry

Top-level SIC	Description	Fraud	Non-fraud
16	Heavy construction	2	2
20	Food	1	1
23	Apparel	6	6
35	Industrial machinery	2	1
37	Transportation equipment	4	4
38	Instrument mfg.	6	7
59	Misc. retail	3	3
60	Depository institution	30	25
61	Non-depository credit institution	1	2
62	Security & commodity brokers	5	5
64	Insurance agents	3	3
67	Holding	4	5
73	Business services	32	35
87	Engineering	6	6

Panel C: Observations by year

Year	Observations
2006	3
2007	40
2008	64
2009	52
2010	28
2011	18
2012	17
2013	8

consists of 114 fraudulent quarters from 31 companies, and the non-fraudulent sample contains 116 quarters from 31 companies. I selected control variables that may have influenced the level of similarity or were important in prior studies. Appendix A contains a list of these variables and descriptions of each.

The similarity scores (SIMSCORE) are separated by part of the call. In relation to other studies that use cosine similarity measurements, Brown and Tucker (2011) report a mean similarity score of 0.845 when comparing current-period MD&As to prior period MD&As. The similarities of the prepared portion of the calls in the current studies is lower than this figure, likely due to the spoken nature of the prepared part of the call. Lee (2016) reports a mean similarity of 0.872 when comparing CEO speech from the Q&A portion of conference to the CEO speech in the prepared portion of the call. The similarity of the Q&A and MD&A in the current study are considerably lower; however, Lee uses function words (e.g. *a, an, the, or*), which makes the similarity scores between these studies difficult to compare.

I include MD&A length (measured in number of words; MDA_LEN) in the event that a longer MD&A section represents a more complex story and therefore requires the executives to maintain more consistency in their disclosures. SIMSCORE correlates with MDA_LEN, which supports the inclusion of this control variable in my regression and classification models.

Regression Model

Because of the considerable difference in means of SIMSCORE between parts of the calls and the interaction of analysts in the Q&A, I ran separate regressions for each part of the call, and for CEO and CFO, resulting in four regressions. I use a random-effects model to estimate the time-invariant coefficient of FRAUD and its effect on the similarity between the two documents. This model is estimated using cluster-robust standard errors, with clustering at the firm level. LOSS is dummy-coded, with 0 representing net income equal to or greater than \$0, and 1 for net income less than \$0. QUARTER is a dummy variable to control for the effects that reporting in a specific period may cause. The number of observations between CEO and CFO regressions vary based on participation in the calls, with 199 CEO observations and 178 CFO observations. The results of these regressions are in Table 10.3 (CEO) and Table 10.4 (CFO).

For CEO language, there were no differences in similarity scores between the CEOs of fraudulent companies and the CEOs of non-fraudulent companies for both the prepared remarks and the Q&A remarks. For CFO language, the CFOs in fraudulent firms had significantly higher language similarity between their prepared remarks and the same-period MD&A than non-fraudulent CFOs.

Table 10.3 CEO language
similarity regressions

	Prepared	Q&A
(Intercept)	2.6203	0.7887
	(2.4010)	(0.7000)
fraud2	0.0236	0.0022
	(0.0344)	(0.0166)
loss	−0.0228	−0.0011
	(0.0276)	(0.0189)
log(atq)	0.0169	0.0066
	(0.0112)	(0.0060)
Quarter_2	0.0294	**0.0230***
	(0.0203)	(0.0111)
Quarter_3	0.0154	−0.0101
	(0.0253)	(0.0121)
Quarter_4	**0.0640***	0.0239
	(0.0246)	(0.0146)
log(age)	−0.4081	−0.1038
	(0.4161)	(0.1148)
soft.assets	**−0.1883***	−0.0379
	(0.0865)	(0.0746)
sch.rec	0.4037	0.0665
	(0.3636)	(0.1535)
sch.inv	0.1852	0.1161
	(0.3610)	(0.1448)
sch.roa	−0.1045	−0.2306
	(0.2175)	(0.1376)
capmkt	−0.0142	−0.0165
	(0.0222)	(0.0097)
meetbeat	**0.0322***	0.0120
	(0.0138)	(0.0096)
R^2	0.2447	0.2061
Adj. R^2	0.1255	0.0807
Num. obs.	199	199

$^{***}p < 0.001$, $^{**}p < 0.01$, $^{*}p < 0.05$

Classification Accuracy

To test for the predictive ability of the variables in the regression (RQ1), I used several different classification algorithms in Weka.[2] The calls were again separated by call segment and speaker, and the regression variables, including SIMSCORE, were used as the feature sets. Each algorithm attempts to predict fraud using the variables included in the regressions. The results are reported in Table 10.5. I use

---

[2] Open-source machine learning software, available at www.cs.waikato.ac.nz/ml/weka/

Table 10.4 CFO language similarity regressions

	Prepared	Q&A
(Intercept)	2.3949	0.4575
	(2.1744)	(0.7390)
fraud2	**0.0996***	0.0048
	(0.0415)	(0.0177)
loss	−0.0305	−0.0135
	(0.0195)	(0.0138)
log(atq)	0.0015	−0.0058
	(0.0165)	(0.0069)
Quarter_2	−0.0171	0.0174
	(0.0214)	(0.0179)
Quarter_3	−0.0119	0.0219
	(0.0246)	(0.0156)
Quarter_4	−0.0579	0.0104
	(0.0321)	(0.0151)
log(age)	−0.4500	−0.0922
	(0.3827)	(0.1265)
soft.assets	0.0310	0.0526
	(0.1077)	(0.0490)
sch.rec	0.4076	0.0558
	(0.2197)	(0.0892)
sch.inv	0.4309	−0.0319
	(0.2444)	(0.0967)
sch.roa	−0.2405	0.1287
	(0.1921)	(0.1600)
capmkt	0.0157	−0.0043
	(0.0193)	(0.0152)
meetbeat	0.0266	**0.0259****
	(0.0135)	(0.0084)
R^2	0.2224	0.1648
Adj. R^2	0.0824	0.0144
Num. obs.	178	178

***$p < 0.001$, **$p < 0.01$, *$p < 0.05$

Table 10.5 Classification accuracies

	Prepared	QA
Logistic	56.1%	58.7%
Naïve Bayes	62.6%	53.5%
C4.5	71.3%	53.5%
SVM	57.4%	58.7%

multiple algorithms because of the possibility that any single method may vary considerably in its performance (Gaganis 2009). Each algorithm was tested using 10-fold cross validation.

In the two models that performed better on the prepared portion (Naïve Bayes and C4.5), the results were significantly higher than chance. The decision tree generated in this C4.5 model shows that similarity score is a very useful variable in discriminating between the two groups. At each level of the tree, the algorithm selects the variable that is best able to reduce uncertainty. Variables near the top of the tree reduce the uncertainty the most. In the prepared remarks, SIMSCORE appears near the top of the tree (see Fig. 10.2).

The logistic regressions and support vector machines (SVM) showed little difference between the parts of calls. These results do provide some insight into RQ2. The prepared remarks, at the very least, perform as well as the Q&A, and may perform better. These results were similar to other studies that tested classification accuracy using language-based models: Larcker and Zakolyukina (2012) achieve accuracy between 56% and 66%, and Humpherys et al. (2011) achieve between 58% and 67%.

Discussion

In this study, I explored the possibility that executives in fraudulent companies would use more similar language between their earnings call and MD&A than non-fraudulent companies. To test this, I collected earnings calls and MD&A texts from fraudulent and non-fraudulent firms. I used similarity scores to measure marginal information disclosure between venues, where lower scores indicate more information. The CFOs of fraudulent firms had significantly greater similarity scores than their non-fraudulent peers during the prepared remarks. This may be the result of attempting to maintain a more consistent story across disclosure venues. These differences did not show up for CEOs, which may be caused by a lower amount of involvement in preparing the MD&A than the CFO (Mayew et al. 2016).

There were no statistically significant differences in similarity between fraudulent and non-fraudulent companies in the Q&A portion of the call, although the

Fig. 10.2 C4.5 Tree. This image shows the importance of SIMSCORE in the decision tree. Values near the top reduce uncertainty the most. The full tree was seven levels; it is trimmed for brevity here

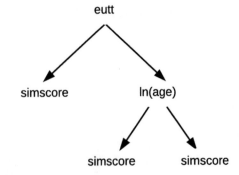

fraud coefficients were greater than zero. Analysts generally control the dialog in this setting, which may limit the ability of managers to use strategic behavior.

This study has several limitations. While AAERs can reasonably establish the ground truth of fraud and non-fraud, it is often unclear if those preparing and presenting the information in the earnings calls and financial statements are aware of the fraud. The size of the sample (230 total firm-quarters) may not provide enough statistical power to detect effects if they are there. However, this number is similar to other fraud research based on AAER data. Humpherys et al. (2011) used 101 fraudulent MD&As from 10-Ks, and Larcker and Zakolyukina (2012) used conference calls from 274 fraudulent firm-quarters. The use of tf-idf and cosine similarity as a measurement tool may be too coarse to capture the meaningful modifications between documents. Deeper semantic analysis may perform better by identifying topical subsets of financial statements and earnings calls. Lastly, it does not consider information contained in the earnings press release.

Future research in this area should investigate the role of financial analysts in uncovering (or facilitating) fraud through their interaction with executives in the earnings calls. It is known that executives choose to speak with analysts who are more favorable to the company, and they may select those who minimize their risk of being caught. Analysts drive the conversation in the Q&A, and therefore have the potential to mitigate strategic information disclosure by managers.

Acknowledgement   We are grateful to the Army Research Office for funding much of the work reported in this book under Grant W911NF-16-1-0342.

Funding Disclosure   This research was funded in part by a grant from the Center for Leadership Ethics at the University of Arizona.

References

Association of Certified Fraud Examiners. (2013, August). *The SEC's newest tool is headed to the front*. Association of Certified Fraud Examiners. http://www.acfe.com/fraud-examiner. aspx?id=4294979300

Association of Certified Fraud Examiners. (2014). *Report to the nations on occupational fraud and abuse: 2014 global fraud study*. Association of Certified Fraud Examiners.

Bloomfield, R. (2012). Discussion of detecting deceptive discussions in conference calls. *Journal of Accounting Research, 50*(2), 541–552.

Brown, S. V., & Tucker, J. W. (2011). Large-sample evidence on firms' year-over-year MD&A modifications. *Journal of Accounting Research, 49*(2), 309–346.

Buller, D. B., & Burgoon, J. K. (1996). Interpersonal deception theory. *Communication Theory, 6*(3), 203–242.

Cacioppo, J. T., & Petty, R. E. (1989). Effects of message repetition on argument processing, recall, and persuasion. *Basic and Applied Social Psychology, 10*(1), 3–12.

Cecchini, M., Aytug, H., Koehler, G. J., & Pathak, P. (2010). Detecting management fraud in public companies. *Management Science, 56*(7), 1146–1160.

Davis, A. K., & Tama-Sweet, I. (2012). Managers' use of language across alternative disclosure outlets: Earnings press releases versus MD&A. *Contemporary Accounting Research, 29*(3), 804–837.

Dechow, P. M., Ge, W., Larson, C. R., & Sloan, R. G. (2011). Predicting material accounting misstatements. *Contemporary Accounting Research, 28*(1), 17–82.

Dyck, A., Morse, A., & Zingales, L. (2010). Who blows the whistle on corporate fraud? *The Journal of Finance, 65*(6), 2213–2253.

Dyck, A., Morse, A., & Zingales, L. (2013). How pervasive is corporate fraud? *Rotman School of Management Working Paper, 2222608*.

Gaganis, C. (2009). Classification techniques for the identification of falsified financial statements: A comparative analysis. *Intelligent Systems in Accounting, Finance and Management, 16*(3), 207–229.

Godfrey, J., Mather, P., & Ramsay, A. (2003). Earnings and impression management in financial reports: The case of CEO changes. *Abacus, 39*(1), 95–123.

Hoberg, G., & Lewis, C. (2014). Do fraudulent firms strategically manage disclosure? *SSRN, 2298302*. http://ssrn.com/abstract=2298302

Humpherys, S. L., Moffitt, K. C., Burns, M. B., Burgoon, J. K., & Felix, W. F. (2011). Identification of fraudulent financial statements using linguistic credibility analysis. *Decision Support Systems, 50*(3), 585–594.

Karpoff, J. M., Lee, D. S., & Martin, G. S. (2008). The cost to firms of cooking the books. *Journal of Financial and Quantitative Analysis, 43*(3), 581–611.

Karpoff, J. M., Koester, A., Lee, D. S., & Martin, G. S. (2012). A critical analysis of databases used in financial misconduct research. *Mays Business School Research Paper, 2012–73*, 2012–11.

Larcker, D. F., & Zakolyukina, A. A. (2012). Detecting deceptive discussions in conference calls. *Journal of Accounting Research, 50*(2), 495–540.

Leary, M. R., & Kowalski, R. M. (1990). Impression management: A literature review and two-component model. *Psychological Bulletin, 107*(1), 34.

Lee, J. (2016). Can investors detect managers' lack of spontaneity? Adherence to predetermined scripts during earnings conference calls. *The Accounting Review, 91*(1), 229–250.

Lo, K., & Rogo, R. (2014). Earnings management and annual report readability. *Working Paper*.

Loughran, T., & McDonald, B. (2011). When is a liability not a liability? Textual analysis, dictionaries, and 10-Ks. *The Journal of Finance, 66*(1), 35–65.

Manning, C. D., Raghavan, P., & Schutze, H. (2008). *Introduction to information retrieval*. Cambridge: Cambridge University Press.

Matsumoto, D., Pronk, M., & Roelofsen, E. (2011). What makes conference calls useful? The information content of managers' presentations and analysts' discussion sessions. *The Accounting Review, 86*(4), 1383–1414.

Mayew, W. J., Sethuraman, M., & Venkatachalam, M. (2016). *"Casting" a doubt: Informational role of analyst participation during earnings conference calls*.

Merkl-Davies, D. M., & Brennan, N. (2007). Discretionary disclosure strategies in corporate narratives: Incremental information or impression management? *Journal of Accounting Literature, 26*, 116–196.

Merkl-Davies, D. M., Brennan, N. M., & McLeay, S. J. (2011). Impression management and retrospective sense-making in corporate narratives: A social psychology perspective. *Accounting, Auditing & Accountability Journal, 24*(3), 315–344.

Petty, R. E., & Cacioppo, J. T. (1979). Effects of message repetition and position on cognitive response, recall, and persuasion. *Journal of Personality and Social Psychology, 37*(1), 97–109.

Purda, L., & Skillicorn, D. (2015). Accounting variables, deception, and a bag of words: Assessing the tools of fraud detection. *Contemporary Accounting Research, 32*, 1193–1223. https://doi.org/10.1111/1911-3846.12089.

Walczyk, J. J., Mahoney, K. T., Doverspike, D., & Griffith-Ross, D. A. (2009). Cognitive lie detection: Response time and consistency of answers as cues to deception. *Journal of Business and Psychology, 24*(1), 33–49.

Zhou, L., & Zhang, D. (2008). Following linguistic footprints: Automatic deception detection in online communication. *Communications of the ACM, 51*(9), 119–122.

Chapter 11
Cultural Influence on Deceptive Communication

Judee K. Burgoon, Dimitris Metaxas, Jay F. Nunamaker Jr,
and Saiying (Tina) Ge

Cultural Influence on Deceptive Communication

A major impetus of the current project was to investigate the under-explored effects
of culture on deception. A wealth of U.S. State Department bulletins and traveler
guidebooks present assumed norms and taboos associated with countries of interest,
and popular magazine articles present anecdotes highlighting appropriate or inap-
propriate conduct. Yet few rigorous empirical investigations have documented
actual communicative behavior, much less deceptive communication behavior, that
can be linked to culture. One exception is the survey completed by the Global
Deception Research Team (2006) that queried respondents from 75 countries on
their perceptions of deceptive behavior but did not measure actual behavior. Another
exception is a comprehensive appraisal of cross-cultural differences from 44 countries in judging actual behavior—in this case, judgments of honesty based on pho-
tographs of smiling—but only one actual enacted behavior was reviewed.

Part of the difficulty in connecting culture to deception stems from the lack of
consensus on whether culture should be defined at the individual or group level.
One of the reasons that nations are typically used as a proxy for culture is that people in the same nation or region of the world are more likely to share language,
education, economy, religions and political systems. Generally speaking, people in
the same nation share more in common with one another than with people from different nations. Our approach to this conundrum of how best to instantiate culture
has been to take a two-pronged approach. As detailed in our chapter on "Culture and

J. K. Burgoon (✉) · J. F. NunamakerJr · S. (Tina) Ge
University of Arizona, Tucson, AZ, USA
e-mail: judee@email.arizona.edu; jnunamaker@cmi.arizona.edu; ge1@email.arizona.edu

D. Metaxas
Rutgers University, New Brunswick, NJ, USA
e-mail: dnm@cs.rutgers.edu

© Springer Nature Switzerland AG 2021
V. S. Subrahmanian et al. (eds.), *Detecting Trust and Deception in Group
Interaction*, Terrorism, Security, and Computation,
https://doi.org/10.1007/978-3-030-54383-9_11

Deception," one prong is to incorporate multiple group-level indicators of culture. These include one's first language, country of birth, and current nationality. The other prong is to allow culture to be defined not by group categories but by individuals' own personal cultural identities. Group categories fail to take into account the degree of mobility in and out of some countries, and the varying degrees of assimilation that takes place. Consequently, we strove to identify dimensions rather than categories that separate cultures and to assess how individuals self-identify. We explore how these various designations do or do not add clarity to understanding cross-cultural deception.

Five questions framed the current chapter:

1. How should culture be measured?
2. Does culture affect a deceiver's motivation, cognitive difficulty, and/or nervousness?
3. Does culture shape behavioral patterns relevant to deception?
4. Does culture affect deceivers' success?
5. What facets of behavior most influence deceptive success?

The SCAN project was not our first foray into investigating the influence of culture on deception. A large-scale experiment funded by the Defense Academy for Credibility Assessment (Burgoon et al. 2008a, b) was undertaken to discover rapid, reliable, and noninvasive technologies and techniques for assessing credibility and malintent. The experiment simulated the types of interviews conducted for rapid credibility assessment and pre-employment screening (see also Meservy et al. 2005). A key consideration was the extent to which deceptive message production and its detection transcend culture, in other words, are universal, or knowledge claims are culture-specific, in other words, must be adjusted by culture. Even in the case of human physiology, which one might assume registers effects equally (Levine et al., 2001), Burgoon et al. (2008a) noted, "Unknown is whether these physiological changes are universal or culture-specific. For example, it is unknown whether fear of detection is triggered by the same circumstances or experienced in the same way in cultures where harsh treatment of prisoners and dissenters has been commonplace as compared to more pacifistic cultures" (p. 7).

Consequently, this investigation, which served as foundational for the SCAN project, measured several culturally relevant concepts and tested the relationships among the self-report dimensions, nationality, deception and deception detection. Below we summarize the experimental procedures, self-report measures, and behavioral measures, which are classified according to their potential for cultural variability. We then proceed to answer the five guiding questions.

Rapid Credibility Assessment Method

Overview

In overview, a laboratory experiment was conducted during which interviewees, who represented diverse cultural backgrounds, responded first to 13 close-ended questions (e.g., yes/no) then 12 open-ended questions posed by professional interviewers. All interviewees gave truthful answers half the time and deceptive answers half the time in a counterbalanced predetermined sequence. Those randomly assigned to the truth-first sequence answered three questions truthfully, three deceptively, three truthfully and three deceptively. Those assigned to the deception-first sequence began with three deceptive responses then alternating between truth and deception for each block of three questions. During the interview, three high-speed digital cameras recorded kinesic, proxemic, and vocalic behavior for later behavioral analysis, and a Laser Doppler Vibrometry (LDV) system collected cardiorespiratory data. For each question, interviewees recorded their degree of honesty and interviewers indicated how honest they perceived the responses to be. Prior to and following the interview, interviewees completed several self-report instruments that measured their cultural orientations, nationality, other demographics, social skills, levels of motivation, and perceptions of the interview.

Participants

Participants (N = 220) were solicited in Pima County, Arizona through a variety of mechanisms to maximize cultural diversity. Letters were sent out to all the international student associations through the University of Arizona Office of International Student Programs. These letters invited students, families and friends to gain interviewing experience and feedback, to receive monetary incentives and to receive tips on successful interviewing in the U.S. in exchange for participation in a study entitled, "Credibility During Interviews: Individual and Cultural Variability in Communication During Interviews." Specifically, the advertisements and flyers stated that participants would receive $10 payment for their time, could earn an additional $10 if deemed credible, and would receive interviewing tips following the interview. Advertisements were placed online via Craigslist, in local newspapers, community shoppers, and the campus newspaper, in flyers posted in the local mosque, at the Chinese Cultural Center, at the local community college, and in other locales frequented by international residents. Recording malfunctions, problems with video capture, and other processing difficulties produced smaller sample sizes for some of the automated analyses. For example, the blob, ASM and blink analyses reduced to 201 or 179 usable cases.

Pre-interaction Self-Report Measures

Participants completed several measures beyond basic demographics. Relevant to the current report are the country of origin, race/ethnicity, first language, years residing in an English-speaking country, and cultural-level orientations. Demographically, the sample was 55% male and 45% female. Mean age was 28.9 years (SD = 13.34), with 36% aged 21 and under, 48% aged 22 to 40, and 16% over 40 years of age. By nationality, the majority (65%) listed their country of origin as the U.S. The remaining 35% indicated countries including India (11%), the former Soviet Republics (5%), China (3%), Mexico (3%), various Western European countries (3%), and Jordan, Iran, and Lebanon. The ethnic breakdown of participants was White/non-Hispanic (48%), Asian or Pacific Islander (21%), Hispanic (15%), and Other (15%) (see Table 11.1 for a complete breakdown). One-third (33%) spoke a language other than English as their primary language, and nearly one-fifth (19%) had lived in the U.S. or an English-speaking country 5 years or less.

To measure cultural orientation, participants completed the Singelis, Triandis, Bhawuk, and Gelfand (1995) Horizontal and Vertical Dimensions of Individualism and Collectivism, the Gudykunst and Lee (2003) Interdependent and Independent Self Construal scale, and the Park and Guan (2006) Positive and Negative Face scale. Cultural-level measures are reported at the individual level, where they assess the tendency for members within these groups to hold similar values, beliefs, and practices. In aggregate, they measure group norms.

The Singelis et al. (1995) four-dimensional scale was intended to produce a more nuanced measure of cultural orientation than had been used traditionally by combining the vertical dimension of human relationships with the horizontal dimension. These are generally viewed as the two primary dimensions along which members of a given culture can be arrayed. Crossing the two major dimensions of the degree of

Table 11.1 Nationality and/or ethnicity of experiment participants

Nationality/ethnicity	N	Percent of participants
U.S. Citizen	143	65%
Caucasian	98	45%
Hispanic	21	10%
Black/African American	14	6%
Asian/Pacific Islander	6	3%
International	76	35%
India	25	11%
Other Asian/Pacific Islander	16	7%
Former Soviet Republics	11	5%
Middle East	3	1%
Spanish Speaking Countries	11	5%
Western/Central Europe	8	4%
Other	3	1%
Total	220	100%

individualistic or collectivist orientation with the degree of hierarchy supported in the culture produced four hybrid dimensions. The *HI* dimension reflects perceptions of autonomy in society and equality among its members. Respondents rated how closely they would describe themselves by statements such as "I often do my own thing," "I prefer to be direct and forthright when discussing with people," and "I am a unique individual." The *VI* dimension reflects perceptions of autonomy in society, focus on personal advancement up the social ladder, and acceptance of inequality. Items capturing this dimension include, "It annoys me when other people perform better than I do," "Competition is the law of nature," and "When another person does better than I do, I get tense and aroused." The *HC* dimension reflects perceptions of being part of a larger collective that experiences equality among its members. Typical items include, "The well-being of my co-workers is important to me," "If a relative were in financial difficulty, I would help within my means," and "It is important to maintain harmony within my group." The final dimension, *VC,* reflects perceptions of being part of a larger collective but with acceptance of inequality and social hierarchy. Scale items include, "I would sacrifice an activity that I enjoy very much if my family did not approve of it," "I usually sacrifice my self-interest for the benefit of the group," and "Children should be taught to place duty before pleasure."

Two further measures were selected to capture individual-level cultural orientations. The Gudykunst and Lee (2003) self-construal instrument assesses how individuals perceive themselves relative to others in their group. The independent self-construal dimension reflects a perception that the self is separate and unique from the group. Typical scale items include "My personal identity is important to me," "I take responsibility for my own actions," and "It is important for me to act as an independent person." The interdependent self-construal dimension reflects a perception that the self is interconnected and part of the group. Typical items are "I will sacrifice my self-interest for the benefit of my group," "I maintain harmony in the groups of which I am a member," and "I respect the majority's wishes in groups of which I am a member."

The remaining cultural self-orientation measure was Park and Guan's (2006) Face Scale, which is an extension of Brown and Levinson's (1987) original politeness scale. Face-saving and face-threats have long been thought to be important considerations in communication in Asian cultures, which are often classified as high-context cultures because much of a message's meaning must be extracted from the context. Comparatively, many western cultures are classified as low-context cultures because more of a message's meaning resides in the message itself.

The politeness scale measures positive and negative face. *Positive face* relates to one's need for approval, a sense of self-worth and favorable identity. It was measured by such items as "It is important for me to look good in front of other people," "Maintaining a positive image is important to me," and "Making a good impression is important to me." *Negative face* relates to preserving one's autonomy and freedom from imposition by others; it centers on one's need for freedom and autonomy. It included these items: "I want my privacy respected," "I have clear boundaries for other people," and "I prefer to keep people at a distance."

To achieve data reduction and create a set of parsimonious dimensions to describe our international sample, the self-orientation and face measures were subjected to principal components exploratory factor analysis with varimax rotation. Using the Kaiser criterion (an eigenvalue of 1.0 as the minimum cut-off for factoring), the Scree test as an indication of the best factor solution, and loading criteria of .50 or better for primary loadings and secondary loadings at least .20 below the primary loading, and at least two items per factor, the best solution to emerge that was interpretable and produced the most distinct dimensions was a seven-factor solution. After removal of items with low primary loadings (below .50) or high cross-loadings, the following seven dimensions, accounting for 58% of the variance, were retained:

1. *Horizontal individualism (HI)*: measures orientation toward individual unique-ness, responsibility and action
2. *Horizontal collectivism (HC)*: measures extent to which group harmony takes priority over personal preferences and goals
3. *Vertical individualism (VI)*: measures extent to which individual is competitive and puts self advancement over that of others
4. *Vertical collectivism (VC)*: measures degree to which individual subordinates his or her self-interest to those of the family and superiors
5. *Self positive face*: measures degree to which individual is concerned with own self presentation and favorable image
6. *Self negative face*: measures degree to which individual values own freedom and independence
7. *Other face*: measures extent to which individual is concerned with providing approbation for another person's face and not imposing on the other

The factor structure produced seven dimensions with acceptable Cronbach's coefficient alpha reliabilities (available upon request). Pearson product-moment correlations confirmed that these dimensions were relatively independent of one another (absolute average $r = .14$, range = .008–.322).

The four combinations found by crossing the vertical-horizontal and individualistic-collectivistic orientations were retained, and the face measures fac-tored into three dimensions, two related to own positive and negative face protec-tion, respectively, and protection of another's face as a single dimension.

The descriptive statistics for the seven measures appear in Table 11.2. Inspection of the means reveals that participants most identified themselves as subscribing to being egalitarian (horizontal) and group-oriented (collectivist), followed by being horizontal and individualist. Since collectivism and individualism are thought to be polar opposites, this kind of responding suggests either people see themselves as having elements of each orientation or the measures are not capturing what was intended. The dimension attracting the lowest response is *VC*, further reinforcing the opposite dimensions of *HI* as more characteristic of these participants. The stan-dard deviations for the two measures including verticality also suggest that there is great variability in how hierarchical the cultures represented here are.

Participants can vary substantially in their manner of communication and possi-bly how they deceive. To assess the extent of variability in the sample and to

Table 11.2 Means, standard deviations and scale range for the seven self-report measures of cultural orientation

Measures of cultural orientation	N	Min	Max	Mean	Std. deviation
Vertical individualism	219	1.00	9.00	5.808	1.636
Horizontal individualism	219	3.83	8.33	6.746	.856
Horizontal collectivism	219	3.75	8.50	7.029	.987
Vertical collectivism	219	1.00	8.50	4.822	1.386
Self positive face	218	1.67	7.00	5.747	.930
Self negative face	218	3.4	7.00	5.955	.656
Concern for other's face	218	2.25	7.00	5.325	.897
Valid N (listwise)	218				

incorporate, if necessary, a covariate of social skills, participants were asked to complete an abbreviated version of Riggio's (1986) Social Skills Inventory. The full SSI consists of 120 items that measure six dimensions. Three are related to verbal skills. *Social expressivity* taps into a person's ability to express oneself effectively in a variety of social contexts, *social control* taps into one's ability to manage and adapt one's verbal communication, and *social sensitivity* taps into one's ability to interpret others' verbal communication. The other three dimensions concern non-verbal skill. *Emotional expressivity* reflects one's ability to make one's emotional expressions understood, *emotional control* reflects the ability to mask one's true feelings, and *emotional sensitivity* reflects one's ability to read others' emotional states. To minimize subject fatigue, we used an abbreviated 30-item version of the SSI that has been used successfully in previous research. This version retained five items per dimension. The coefficient alpha reliabilities for the three social dimensions were .76, .69, and .56 and for the three nonverbal dimensions, .50, .69, and .61, respectively.

Interaction Self-Report Measures

Several manipulation checks were conducted to ensure that participants followed instructions After responding to each question in the interview, participants were asked to rate themselves on an 11-point scale from completely untruthful (0) to completely truthful (10). Responses of 6 or above were regarded as truthful; responses of 5 or below were regarded as deceptive. Participants who failed to follow instructions (instructed to lie but responded truthfully or vice versa) had their answers tagged. Those who had four or more tagged responses were disqualified for the analyses. Four was chosen rather than a lower number to allow room for the margin of error inherent in self-report questions. Eighteen participants were disqualified as a result of this manipulation check, leaving 201 valid cases.

Following the interview, participants were asked to rate their level of arousal, motivation, and cognitive difficulty during the interview on a 1 (not at all) to 7 (very high) rating scale. These measures reflect theorized triggers for deceptive displays. The coefficient alpha reliabilities for the 12 items measuring these states were .88 for negative arousal, .85 for cognitive difficulty and .82 for post-interaction reported motivation.

Procedure and Independent Variables

Participants arriving at the research site were greeted by a research assistant who followed a standardized script. Participants were randomly assigned to one of the two sequence conditions (truth-first or lie-first), completed consent forms and completed the pre-interaction self-report measures. The assistant described the upcoming interview and delivered instructions for truth-telling and deceit. See Table 11.3 for interview questions.

Next, participants were ushered into the interview room, which was equipped with three professional grade high-speed (30 fps) digital cameras mounted on tripods, a series of video monitors, an LDV system and a near-infrared camera for

Table 11.3 Interview questions and types

Questions	Content	Verifiability
1. What is the worst job you ever had and why did you dislike it?	Autobiographical judgment	Not verifiable
2. What would you do if your boss gave you credit for someone else's work?	Moral dilemma	Not verifiable
3. Tell me about a time when you thought of stealing something valuable from someone.	Past-oriented narrative	Not verifiable
4. Please tell me everything you did today from leaving your home to arriving for this interview.	Past-oriented narrative	Verifiable
5. What do you consider to be your greatest strengths?	Autobiographical judgment	Not verifiable
6. What else are you going to do today? Who will you see and where will you go?	Future-oriented narrative	Verifiable
7. Think about people that really irritate you. Why do they bother or annoy you?	Autobiographical judgment	Not verifiable
8. What do you plan to do during your next break or vacation?	Future-oriented narrative	Verifiable
9. Remember the room where you arrived for the experiment. Tell me everything about that room and what happened while you were there.	Past-oriented narrative	Verifiable
10. Tell me about a time when you told a serious lie to get out of trouble.	Past-oriented narrative	Not verifiable
11. If you found a wallet containing $1000 and no identification in it, what would you do with it and why?	Moral dilemma	Not verifiable
12. What is the worst restaurant you ever went to? Why did you dislike it?	Autobiographical judgment	Not verifiable

measuring eye behavior (not reported here), and a teleprompter managed by the AV technician. Participants sat approximately 8′ from the interviewer, facing the interviewer and teleprompter. Adjacent to the interviewee was a small table with a computer mouse that the interviewee used to control the on-screen rating form. Following the interview, participants were taken to a separate room to complete post-measures and be debriefed. The floor plan, location of personnel and flow of participants among rooms are shown in Fig. 11.1.

One camera recorded the participant's face and another recorded the full frontal body image. A third camera captured both the interviewer and interviewee in profile. The teleprompter was positioned below one camera. Just before each question, it instructed the interviewee whether to tell the truth or be deceptive. The interviewer was blind to the truth or lie status and to the proportion of truthful or deceptive responses. The two different sequences (truth-first or deception-first) were counterbalanced to control for possible order effects. The assistant explained the purpose of each piece of equipment and introduced the interviewee to the interviewer, after which the interview began.

The interview sequence incorporated a mix of autobiographical, narrative and opinion question types that required lengthy open-ended responses. The autobiographical questions asked about the self and included judgments. The narrative questions (also referred to as episodic memory) prompted longer responses and asked for many details about past events or future plans. The opinion questions

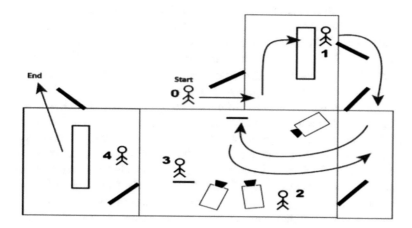

Key:
0 = Subject
1 = Greeter / Pre-tester
2 = AV Tech /Prompter
3 = Interviewer
4 = Debriefer / Post-tester

Fig. 11.1 Assistant stations and flow diagram for experiment rooms

posed moral dilemmas and asked interviewees how they would handle them. The responses requested varied as to whether they asked for facts, conjectures or judgments. They also differed as to whether they were verifiable or not (see Nahari et al. 2014, for the value of verifiability). Interviewers followed the exact wording and sequence of the 12 main questions but were allowed to ask follow-up probes. The interviews followed either a TTT-DDD-TTT-DDD or DDD-TTT-DDD-TTT order, where T represents truth and D represents deception.

Interviewers were five professionals from the intelligence community. They offered feedback on the questions to be asked. After each block of three questions, interviewers rated interviewee truthfulness on the same 0 (not at all truthful) to 10 (very truthful) scale as used by interviewees. The interviewee self-reports and interviewer estimates became the basis for calculating bias and accuracy scores.

Interviews varied in length but most were 20–30 minutes long. Following the interview, interviewees proceeded to the debriefing room where they completed a post-interview questionnaire consisting of post-interaction measures of motivation, arousal, and cognitive difficulty; self-reported social skills; and self-reported cultural orientations. They were then debriefed and paid for their participation.

Physiological and Behavioral Measures

Significant nonverbal behavior research has established that cultures vary on a variety of features that follow cultural dictates called cultural display rules (Buller & Burgoon, 1996). Cultures influence what behaviors should be displayed, which ones are taboo, and what consequences flow from each action (Krys et al. 2016). However, of interest here are emotional and cognitive responses associated with deception. The question of whether physiological responses are universal or not mandates measuring one or more of these indicators. The issue of controllability—how well deceivers control kinesic behaviors (facial expressions, gestures) and vocalic indicators of stress and cognitive difficulty—warrants examination of various voice measures of amplitude, cognitive effort and fluency. Though language itself is part of the definition of culture and ethnicity and so is variable across nation state and geographic boundaries (Mooney & Evans 2019), linguistic features such as syntax and lexical quantity may reflect the influence of deception and so deserve to be examined for any interaction with culture. Table 11.4 shows instruments for collecting physiological and behavioral indicators and the predicted extent of cultural influence.

The possible interactions between culture and deception were instantiated with these indicators:

1. *Eyeblink*—Blinks are thought to be a relatively involuntary reflex controlled by the autonomic nervous system (Fukuda, 2001). Blinks were captured with a JVC GY-DV550 professional grade, high-speed visible spectrum digital camera. Blinks were measured with a computer vision tool, the ASM Face Tracker, which

Table 11.4 Instruments for collecting physiological and behavior indicators

Technologies and behavioral indicators	Consistency across culture
ASM Tracker of eyeblink	High probability unless stimuli being viewed are culture-specific
ASM Tracker of kinesic facial behavior	Low probability
Skin Blob Tracker of kinesic gestural behavior	Low probability
LVA analysis of vocalic stress	High probability
SPLICE linguistic analysis	Low probability

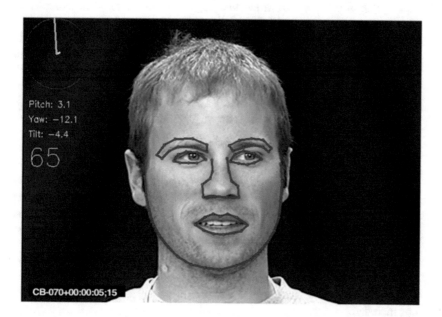

Sample frame showing tracking of the 79 facial landmarks. The circle in the upper left corner depicts the estimated 3D vector of the head pose. Underneath it are the estimated values of the pitch, yaw and tilt angles of the head

Fig. 11.2 Active shape modeling landmarks on the face

is a software suite patented by Rutgers University (PI: Dimitris Metaxas). The ASM Tracker accurately tracks the 2D position of 79 facial landmarks that define the contour of the face, eyes, eyebrows, nose and mouth in real time from a single camera view (see Fig. 11.2). The software uses an Active Shape Model (ASM) together with a novel nonlinear shape subspace method to handle large head rotations, so that tracking is robust and continuous.

2. *Head nods and shakes*—These kinesic behaviors may reflect an interviewee's emotional tenor in the form of communicative expressivity or rigidity, an interviewee's cognitive effort in the form of cognitive fluidity or difficulty, or both (Harrigan & O'Connell, 1996). The same camera and ASM Tracker as used

for blinks were used to capture the pitch (forward and backward nodding), roll (left and right shaking) and yaw (tilt) of the head.

3. *Hand shrugs*—These gestures fall into the kinesic category of illustrator gestures (Ekman & Friesen 1969). Such gestures are thought to signal uncertainty, resignation or detachment from what is being said and were measured with another computer vision technique, the Skin Blob Tracker (SBT), which was conducted on the full-body videos. The SBT first identifies the head, hands and shoulders with ellipses then calculates shapes and velocity of movement of the blobs within and across frames.

4. *Vocal pitch, loudness, LVA stress and cognitive indicators*—The voice is one channel that can be very telling about whether an individual is relaxed and at ease, is nervous and stressed, is thinking hard, or is taking an aggressive stance (Kurohara et al. 2001). Acoustic features were extracted from the same videos as the preceding features after being down-sampled from 48 to 12 kHz. Measures of pitch elevation, hesitations, stress and global fight or flight tendencies were analyzed with Nemesysco Layered Voice Analysis (LVA) 6.50. Additionally, signal processing of acoustic parameters was analyzed with proprietary software and MatLab at the Rome Air Force Research Laboratory. Those results are not reported here.

5. *Nonfluencies*—Speech errors were compiled from transcriptions of the responses to the 12 questions then processed using the hesitation module in the Agent99 Analyzer (A99A), a suite of tools developed by Burgoon and colleagues to parse, annotate and analyze several classes of linguistic features thought to differentiate fictive or deceptive from truthful texts (Burgoon et al. 2008a, b). The hesitation module was a lexicon that included garbled sounds such as "ykn," fillers such as "uh," "um," and "okay," expressions such as "I don't know" and consecutively repeated words such as "I I I don't know." These were summed then standardized according to the total number of words within one of the 12 responses, follow-ups included.

6. *Language lexicon and syntax*—All interviews were transcribed then processed through Agent99Analyzer, which used the open-source Natural Language Toolkit for speech tagging, word tokenization, stemming, word frequency counting, categorization, probabilistic parsing and the like (https://sourceforge.net/projects/nltk/) (Loper & Bird 2005).

Results

Manipulation Checks

Several manipulation checks were conducted to assess whether participants followed instructions to lie or be truthful and experienced hypothesized elevated stress, cognitive difficulty, and attempted behavioral control associated with lying.

Categorical variables were tested with *t*-tests or ANOVAs. Relationships between two interval-level metrics were tested with Pearson product-moment correlations.

A 12 (questions) × 2 (condition: truth-first or deception-first) repeated-measures analysis of variance of truthfulness ratings using Huynh-Feldt corrected degrees of freedom due to violation of sphericity assumptions produced a significant question by veracity order interaction, F (11, 5379) = 562.42, p < .0001, partial η^2 = .72. Participants varied their truthfulness by condition and question, as instructed. Those in the truth-first order averaged a rating of 8.95 when responding truthfully and 2.56 when deceiving. Those in the deception-first order averaged 9.63 when responding truthfully and 2.45 when deceiving. Figure 11.3 shows the participants' average self-reported truthfulness per question on the 11-point scale. Those in the truth-first condition answered questions 1, 2, 3, 7, 8, and 9 truthfully and the remainder deceptively. Those in the deception-first condition answered questions 4, 5, 6, 10, 11 and 12 truthfully and the remaining questions deceptively.

How Should Culture Be Measured?

Country of origin, ethnicity and language The traditional way of representing culture has been to use one's birthplace, current country of residence, ethnicity or some combination thereof. However, when participation is strictly voluntary and not controlled as a stratified sample, as was the case in this investigation, the distribution can be seriously uneven. In the current investigation, 28 different countries were represented, some with only a single individual, making any kind of generalization by country of origin impossible. A more condensed distribution grouped by region and with ethnicity incorporated still produced a highly imbalanced sample, with some regions and ethnicities with very few representatives. Even a division by native language produced a lopsided distribution (67% English as first language, 33% another language), as did number of years residing in an English-speaking country (67% reported 3 years or less). By default, culture as measured by self-reported cultural orientation became the most evenly distributed method of reflecting culture. However, to gain more insight, we crossed the dimensions with a compressed classification by nationality and ethnicity. The categories are shown in Table 11.5; see also Fig. 11.4.

Self-orientation One-way ANOVAs with nationality/ethnicity as the independent variable were conducted on these dependent measures: self-reported cultural dimensions, social skills dimensions that measured communication skills, pre-interaction goals and motivations, and post-interaction motivation.

Three of the four cultural orientation dimensions showed significant differences across the groupings. Hispanics and those from former Soviet Republics scored highest on *horizontal collectivism (HC;* placing higher importance on group harmony than own needs) and Europeans scored lowest, although still above the scale's

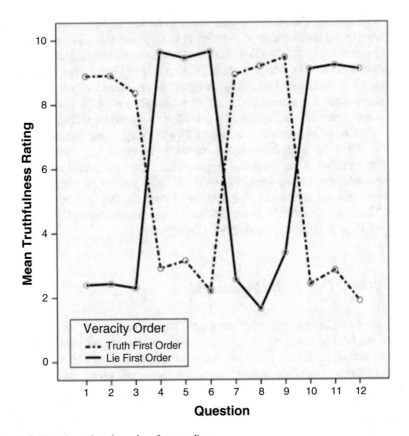

Fig. 11.3 Veracity ratings by order of responding

midpoint. However, the large standard deviation for the latter group indicates that they were not homogeneous in their orientation. On *VC* (hierarchically oriented, with the individual subordinated to the group), Indians and other Asians scored much higher than U.S. Caucasians and other Europeans. On *VI* (status important but emphasis on self-advancement), Indians and Europeans were the two groups that scored highest, whereas Hispanics scored the lowest. *HI* did not differ across groupings. To aid in visualization of these results, the distributions by native country of origin are shown in Fig. 11.4, where countries with 4 or more respondents are included.

Bolstering the results is the finding that the nationalities/ethnicities differed on the four retained social skills dimensions and the three face concerns. U.S. blacks placed higher importance than all the other groupings on social expressivity (knowing what to say and when), and Asians placed the least importance on it. Western Europeans placed greatest importance on social control (being able to control and adapt what one says in various contexts), whereas those from central and eastern Europe placed the least on it. Indians placed most importance on social sensitivity (being a good and accurate listener) and Hispanics placed the least. All the Asian

Table 11.5 Means (and standard deviations) for significant differences by nationality/ethnicity

Variable	F-value	U.S. Caucasian	Hispanic	U.S. Black	Indian	Other Asian	Former USSR	European	Other
Horizontal collectivism	2.38*	6.64	7.20	7.01	7.05	6.88	7.24	6.43	7.01
		(0.93)	(0.86)	(0.78)	(0.78)	(0.89)	(0.80)	(1.39)	(0.67)
Vertical collectivism	3.74**	4.71	5.10	5.48	5.94	5.78	5.61	4.67	5.13
		(1.31)	(1.43)	(1.64)	(1.11)	(1.19)	(1.22)	(1.10)	(1.61)
Vertical individualism	3.69**	5.81	4.89	5.38	6.70	5.59	6.55	6.69	5.80
		(1.58)	(1.72)	(1.76)	(1.34)	(1.83)	(1.33)	(1.44)	(1.05)
Social expressivity	2.65*	3.54	3.63	3.91	3.06	2.87	3.36	3.80	3.63
		(0.91)	(1.02)	(0.79)	(0.93)	(0.69)	(0.94)	(1.14)	(0.87)
Social control	2.15*	3.51	3.57	3.61	3.16	3.04	2.96	3.87	3.45
		(0.85)	(0.84)	(0.82)	(0.78)	(0.87)	(0.54)	(0.71)	(0.90)
Social sensitivity	3.03**	2.84	2.51	2.90	3.32	2.81	3.16	3.11	2.60
		(0.69)	(0.78)	(1.04)	(0.67)	(0.84)	(0.82)	(0.67)	(0.71)

*p < .05, two-tailed. **p < .01, two-tailed

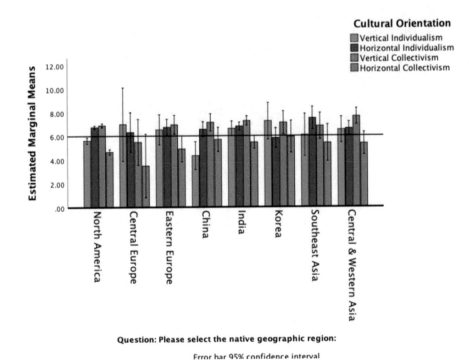

Fig. 11.4 Cultural orientation by geographic region of origin

respondents placed higher importance on negative face concerns than westerners. Positive face concerns and concerns for another's face were also highest among Koreans and Asians other than Chinese (Fig. 11.5).

These results make clear that any findings from a U.S. sample cannot be generalized to other cultures. Even though this sample was collected in the U.S. and the non-U.S. participants might already reflect some degree of assimilation, they still reflected significant differences from native U.S. respondents. Additionally, countries that share similarities on one dimension cannot be assumed to be alike on other dimensions. For example, even though Central and Eastern Europe are fairly similar in having a VI orientation, they are quite different on the two collectivist orientations. Korea and Southeast Asia are very similar in their VC orientation but not on any of the other dimensions.

Finally, it should be apparent that countries do not fit cleanly into one and only one of the four cultural dimension quadrants. The fact that a given nationality or ethnicity group can be described by two or more dimensions is the product of using measures that did not force respondents to choose among the four descriptions. This means cultures cannot be classified by a single dimension and cultures do not fit into uniquely independent categories. Nevertheless, the dimensions do provide some insights into the various geographic groupings. The Chinese are least likely to subscribe to *VI* but are not loathe to hierarchy; the key is that their communal

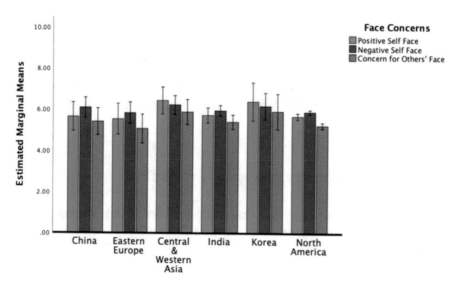

Fig. 11.5 Face concerns by geographic region of origin

orientation means putting the group ahead of the individual. Indians, by contrast, are least likely to see themselves as *HC*s. They, too, subscribe to having a hierarchical society but individuals can place their own self-advancement ahead of the group.

Does Culture Affect a Deceiver's Motivations, Cognitive Difficulty, Arousal and Behavioral Control?

The answer to all of these questions is yes. Theories of deception typically posit that deceivers have various goals or motivations they are attempting to achieve or protect while deceiving, that they experience cognitive difficulty and discomfort in the process of deceiving, and that they attempt to control their behavior. The results here show that those motivations and experiences are not uniform across cultures. Deceivers who hold a *HC* orientation placed more value than others on the goals of putting forward a successful presentation of self and having good interpersonal relationships during the interview. They also reported being more concerned than others about putting forward a positive face and being respectful of others' face. Finally, those with the horizontal orientation were especially concerned with succeeding with their deceit. Conversely, *HI*s placed greater emphasis than others on negative face, that is, protecting their autonomy. They did not place the same emphasis on positive face or protecting others' face. They reported less stress than others during the interview.

*VI*s, like *HC*s, placed greater importance on protecting their own positive face but *less* importance on protecting others' face. They also found the interview more cognitively challenging than did others. *VCs* also rated self-presentation as an important goal. They reported a modest correlation between their cultural orientation and greater effectiveness in controlling their nonverbal behavior during the interview. These correlations are all presented in Table 11.6.

In short, culture influenced what was valued as interaction goals, how stressful and cognitively difficult the interview was, and how *effectively* they controlled their nonverbal behavior.

Does Culture Affect Communication Patterns?

Analyses were conducted on eye blinks, head nods and shakes, gestural shrugs, vocal indicators related to stress and cognitive effort, nonfluencies, and lexical and syntactic linguistic features.

Eye Blinks

The average duration of blinks differed by question and, when grouped by blocks of three questions, by block. Blinks became more rapid but of shorter duration over time. By culture, HCs blinked more during deception but not truth. Otherwise, there were no main effects or interactions with deception order.

Table 11.6 Correlations among cultural orientations, pre-interaction goals, cognitive difficulty and stress while interacting, and perceived success at behavioral control

	Vertical collectivism	Vertical individualism	Horizontal individualism	Horizontal collectivism
Concern for self-positive face	0.168*	0.279**	−0.010	0.360**
Concern for self-negative face	−0.098	−0.021	0.556**	0.141*
Concern for other's face	0.09	−0.217**	−0.012	0.225**
Importance of succeeding	0.053	0.061	0.132	0.142*
Interpersonal relationship goals	0.128	0.012	0.133*	0.362**
Self-presentation goals	0.210**	0.071	0.057	0.243**
Cognitive difficulty	0.094	0.200**	−0.073	−0.055
Stress	−0.101	0.069	−0.132	−0.147*
Attempted control of nonverbal behavior	0.046	0.028	0.062	0.080
Effective control of nonverbal behavior	0.146*	0.005	0.062	0.089

*p < .05, two-tailed. **p < .01, two-tailed

Hand Shrugs

Disregarding culture for the moment, those who began answering questions with lies shrugged more throughout the interview than those who began by telling the truth. This pattern of shrugging across the 12 questions is interesting. As shown in Fig. 11.6, those who begin with deception do more shrugging than truth-tellers during their first three questions, as expected. But then when they switch to truth, they continue to do more shrugging than their counterparts who are now deceiving. This difference continues across all 12 questions. It reveals that whatever pattern interviewees display at the beginning tends to dictate their continued nonverbal behavior, although all interviewees increase shrugging over time. Had measurement been done on a question like Question 1 or 10, the difference would have been slight; on Questions 10, 11 and 12 (i.e., the end of the interview), differences would have been much larger but in the opposite direction of the hypothesis. This illustrates how sensitive these measures are to the type of question being asked and the timing of measurement.

By cultural orientation, *HIs* shrugged *more* during deception, whereas *VIs* shrugged *less* during both deception and truth, in other words, *VIs* were generally less expressive. This would be applicable to countries like India and those in western Europe. Because more shrugs might signal deception among *HIs* but signal truth among *VCs*, it is an unreliable deception indicator. Table 11.7 illustrates correlations between average amount of shrugging by the right hand, calculated by duration and by rate per frame, during truth and deception, for each cultural dimension. Results on the left hand are comparable.

Vocal Pitch, Stress and Cognitive Effort

Analyzed with the Nemesysco tool were F_{main} (the average fundamental frequency), SPT (a measure of emotional level derived from the high frequency range), SPJ (a measure of cognitive activation derived from the low frequency range), APJ a measure of cognitive work based on the average range of low frequencies), Say-or-Stop (changes in SPT or SPJ), JQ (global stress level based on uniformity across low frequencies) and Lie Stress (significant frequencies in the spectrum). Those who began with lies had elevated pitch and it remained elevated although varying by question. Those who began with truth had a more erratic pattern but tended to have higher pitch when shifting to deception. None of the patterns were affected by cultural orientation.

Those higher on *VC* tended to show less emotionality but more thinking (more pauses) in their voice than those low on *VC*. Emotionality and "thinking level" did not differ with veracity. *VCs* tended to show less stress and to reduce their indications of stress during the middle of the interview. Those who began with truth showed more indications of stress and specifically lie stress, and generally maintained a higher level throughout, than those beginning with deception. *VIs* showed

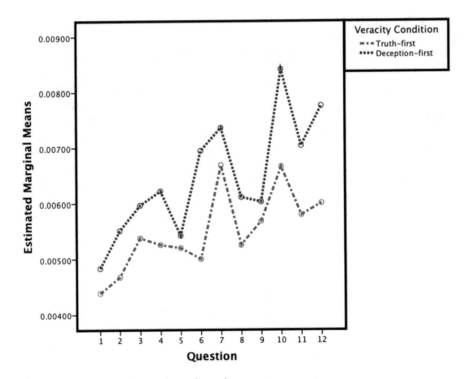

Fig. 11.6 Hand shrugs by question and veracity

Table 11.7 Correlations of hand shrugs during truth and deception

	Horizontal collectivism	Horizontal individualism	Vertical collectivism	Vertical individualism
Average shrugs duration, deceiver	−0.015	0.006	−0.039	0.016
Average shrugs per frame, deceiver	−0.026	0.154*	−0.186*	−0.250**
Average shrugs duration, truth-teller	0.016	−0.037	0.014	0.022
Average shrugs per frame, truth-teller	0.052	0.137	−0.105	−0.272**

*$p < .05$, two-tailed. **$p < .01$, two-tailed

less global stress (fight or flight) across the entire interview. Although *HIs* varied in their lie stress across questions, the correlations were modest.

Those higher on *HC* exerted more cognitive effort especially during the first part of the interview. Those who began with truth also showed more effort than those who began with deception. These patterns would be most true of Koreans and least true of central Europeans.

Lexical and Syntactic Features

A percentage of the linguistic features differed by question and by block as a result of question type being asked. Four of relevance to stress and cognitive effort were number of words, words per sentence, nonfluencies, and fillers (interjections like "um" and "ahh"). Deceivers are expected to have briefer messages than truthtellers. Here, word count interacted with question and blocks of questions. Once past the first block of questions, interviewees had a larger word count when telling the truth than lying. Similarly, interviewees were expected to use syntactically more complex language when telling the truth than lying. That was the case with these interviewees. In addition to the syntactic complexity of sentences varying by question and block, truthful responses were more complex than deceptive ones (see Fig. 11.7). Additionally, lying elicited more nonfluencies than truth-telling, as shown in Fig. 11.8. Results for fillers (words like "um" and "ahh") followed the same pattern.

Culture exerted very little influence on language variability during deception. As one exception, *HC* and *VC* used fewer words in response to each of the 12 questions, but this was probably due to not being native English speakers.

In sum, culture did have some influence on the behaviors being exhibited by interviewees when telling the truth versus deceiving, enough so that ignoring culture when analyzing deception would be a mistake. Although many of the effects were modest, it would still be advisable to have baseline data in a given country or culture against which to make comparisons during possible deception.

Does Culture Affect Deceivers' Success?

Self-perceived success One marker of success is how interviewees rate themselves in being judged. By dimension, all but VIs thought they were successful. Additionally, those who experienced more stress and cognitive difficulty were less likely to rate themselves as successful, whereas those who worked harder and thought they had better nonverbal control thought they were more successful.

Source credibility Another way to measure deceiver success is to have interviewers rate interviewee credibility. To the extent that interviewer ratings vary by interviewee veracity, they indicate that interviewers are accurately detecting interviewee deception. As shown in Fig. 11.9, where credibility ratings are averaged within blocks of questions, deception-first interviewers initially recognized when interviewers lied or told the truth in blocks one and three interviewees answering truthfully were judged as more credible than those answering deceptively. However, in later blocks it became more difficult for interviewers to differentiate between truthful and deceptive answers, making the comparisons to those in the other condition as well as to self when changing own veracity less clear, stark comparisons.

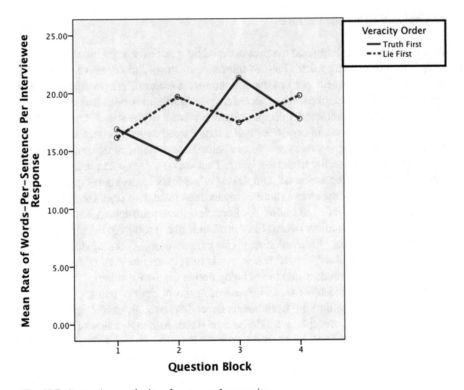

Fig. 11.7 Syntactic complexity of sentences by veracity

Importantly, these ratings were not affected by cultural orientations. Put differently, interviewers' ratings were not biased by the cultural background of who they were rating. Their judgments were highly accurate and unbiased.

Correct judgments by interviewers A final measure of interviewee success is interviewer accuracy in detecting lies and truths. Interviewer estimates of truthfulness were subtracted from interviewees' reports of their own truthfulness. The smaller the difference, the better. As shown in Fig. 11.10, interviewers were accurate in distinguishing lies from truths, meaning interviewees did not escape detection. Additionally, cultural dimensions did not influence interviewer accuracy.

Summary

A major impetus of the current project was to delve more deeply into how culture affects deception and its detection. Five questions were addressed.

The first asked, *How should culture be measured?* The huge variety of nationalities and unevenness in sample size from each mitigated against culture being defined by nationality. Although categories of ethnicity are fewer, unevenness in their

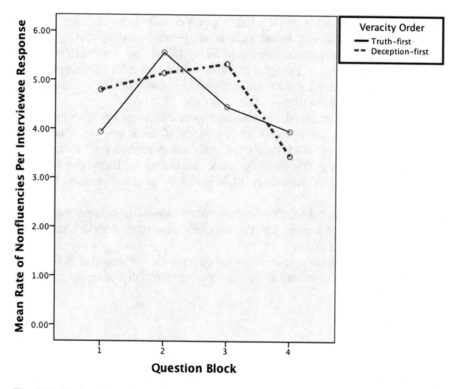

Fig. 11.8 Blocks of three interview questions by veracity order of responding

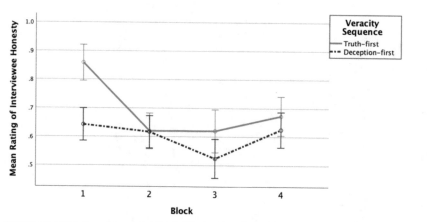

Covariates appearing in the model are evaluated at the following values: Gender = 1.46, Age = 29.14, Is English your first language? = 1.33, HorizontalColl = 6.7980, HorizontalInd = 6.9812, VerticalColl = 5.0331, OtherFace = 5.2530, SelfPosFace = 5.6898, SelfNegFace = 5.9136, VerticalInd = 5.7610

Error bars: 95% CI

Fig. 11.9 Interviewer ratings of interviewee credibility, by blocks of questions

distributions also argued against it being an appropriate definition of culture. Guided by scholars of intercultural communication, we opted for having participants use their own personal orientation. An exploratory factor analysis identified seven dimensions that were used throughout to assess the extent to which deception perceptions, behaviors and detection are universal or culture-specific. As might have been expected, the results supported some of each.

The second question asked, *Does culture affect a deceiver's motivation, cognitive difficulty, and/or nervousness?* The answer in all cases was yes. Participants varied in the degree to which they were motivated to achieve their interpersonal goals, experienced cognitive difficulty while conducting the interviews and were nervous. Thus, observed indicators might differ by culture, unrelated to being deceptive.

The third question asked, *Does culture shape behavioral patterns relevant to deception?* Again, the answer was yes, although more often than not, behavioral patterns transcended culture.

The fourth question asked, *Does culture affect deceivers' success?* In this case, the answer was no. Regardless of the measure, success in detecting deception was not detected.

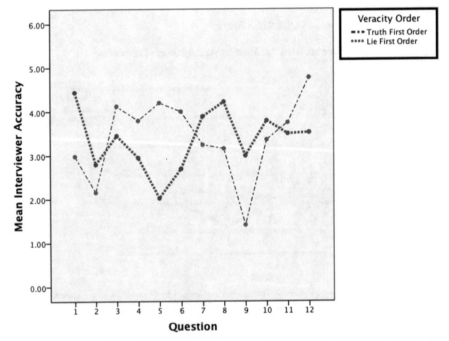

Note: Higher scores represent higher INaccuracy. Accuracy is higher when interviewer ratings are least discrepant from interviewee self-reported truthfulness.

Fig. 11.10 Interviewer accuracy across 12 interview questions

The final question asked, *What facets of behavior most influence deceptive success?* Several behavior measures predicted deception. But their impact was not mediated by culture. More often than not, cultural orientation did not mediate effects.

Can we ignore culture? It appears not. But the impact is nuanced and must be identified for its impact.

Acknowledgement   We are grateful to the Army Research Office for funding much of the work reported in this book under Grant W911NF-16-1-0342. Judee Burgoon and Jay Nunamaker are affiliated with Discern Science International, a for-profit entity that develops systems for credibility assessment.

References

Brown, P. & Levinson, S. C. (1987). *Politeness: Some universals in language usage* (Vol. 4). Cambridge University Press.

Buller, D., & Burgoon, J. (1996). Interpersonal deception theory. *Communication Theory, 6*(3), 203–242.

Burgoon, J. K., Derrick, D. C., Elkins, A. C., Humphreys, S. L., Jensen, M. L., Diller, C. B. R., & Nunamaker, J. F. (2008a). *Potential noncontact tools for rapid credibility assessment from physiological and behavioral cues.* Presented at the IEEE International Carnahan Conference on Security Technology, October 13–16, 2008, Prague, Czech Republic.

Burgoon, J. K., Levine, T., Nunamaker, J. F., Metaxas, D., & Park, H. S. (2008b, August). *Rapid credibility assessment: A report to the counterintelligence field activity (Contract No. H9C 104-07-C-0011).*

Dionisio, D. P., Granholm, E., Hillix, W. A., & Perrine, W. F. (2001). Differentiation of deception using pupillary responses as an index of cognitive processing. *Psychophysiology, 38*, 205–211.

Fukuda, K. (2001). Eye blinks: New indices for the detection of deception. *International Journal of Psychophysiology, 40*, 239–245.

Global Deception Research Team. (2006). A world of lies. *Journal of Cross-Cultural Psychology, 37*(1), 60–74.

Gudykunst, W. B., & Lee, C. M. (2003). Assessing the validity of self construal scales: A response to Levine et al. *Human Communication Research, 29*, 253–274.

Harrigan, J. A., & O'Connell, D. M. (1996). How do you look when feeling anxious? Facial displays of anxiety. *Personality and Individual Differences, 21*, 205–212.

Krys, K., Vauclair, C. M., Capaldi, C. A., Lun, V. M. C., Bond, M. H., Domínguez-Espinosa, A., Torres, C., Lipp, O. V., Manickam, S. S., Zing, C., Antalíková, R., Pavlopoulos, V., Teyssier, J., Hur, T., Hansen, K., et al. (2016). Be careful where you smile: Culture shapes judgments of intelligence and honesty of smiling individuals. *Journal of Nonverbal Behavior, 40*, 101–116.

Kurohara, A., Terai, K., Takeuchi, H., & Umezawa, A. (2001). Respiratory changes during detection of deception: Mechanisms underlying inhibitory breathing in response to critical questions. *Japanese Journal of Physiological Psychology and Psychophysiology, 19*(2), 75–86.

Levine, J. A., Pavlidis, I., & Cooper, M. (2001). The face of fear. *The Lancet, 357*(9270), 1757.

Meservy, T. O., Jensen, M. L., Kruse, J., Burgoon, J. K., & Nunamaker, J. F. (2005). Automatic extraction of deceptive behavioral cues from video. In *Intelligence and security informatics, lecture notes in computer science* (Vol. 3495). Berlin/Heidelberg: Springer.

Mooney, A., & Evans, B. (2019). *Language, society and power* (5th ed.). New York: Routledge.

Nahari, G., Vrij, A., & Fisher, R. P. (2014). Exploiting liars' verbal strategies by examining the verifiability of details. *Legal and Criminological Psychology, 19*, 227–239.

Park, H. S., & Guan, X. (2006). The effects of national culture and face concerns on intention to apologize: A comparison of the USA and China. *Journal of Intercultural Communication Research, 35*, 183–204.

Riggio, R. E. (1986). Assessment of basic social skills. *Journal of Personality and Social Psychology, 51*, 649–660.

Singelis, T. M., Triandis, H. C., Bhawuk, D. P., & Gelfand, M. J. (1995). Horizontal and vertical dimensions of individualism and collectivism: A theoretical and measurement refinement. *Cross-Cultural Research, 29*, 240–275.

Printed in the United States
by Baker & Taylor Publisher Services